GE 42 .I
P9-DSY-469
Helgesen, Leif Magne.
Ice is melting

SOUTH PLAINS COLLEGE LIBRARY
DATE DUE

JUL 18 2016			

PRINTED IN U.S.A.

The Ice is Melting

Ethics in the Arctic

Leif Magne Helgesen, Kim Holmén and Ole Arve Misund (eds.)

The Ice is Melting

Ethics in the Arctic

SOUTH PLAINS COLLEGE LIBRARY

FAGBOKFORLAGET

Copyright © 2015 by
Fagbokforlaget Vigmostad & Bjørke AS
All Rights Reserved

ISBN: 978-82-450-1843-1

Graphic production: John Grieg AS, Bergen
Cover design by Fagbokforlaget

Editors: Leif Magne Helgesen, Kim Holmén and Ole Arve Misund
Photo editor: Eva Therese Jenssen
English translation: Janet Holmén

This book is the result of collaboration between Svalbard Kirke, the Norwegian Polar Institute, and the University Centre in Svalbard.

Cover photo: Samarin glacier calving (Hornsund, Svalbard); Eva Therese Jenssen

Inquiries about this text can be directed to:
Fagbokforlaget
Kanalveien 51
5068 Bergen
Tel.: 55 38 88 00 Fax: 55 38 88 01
e-mail: fagbokforlaget@fagbokforlaget.no
www.fagbokforlaget.no

All rights reserved. No part of this publication may be reproduced, stored in a retrieval system, or transmitted, in any form or by any means, electronic, mechanical, photocopying, recording, or otherwise, without the prior written permission of the publisher.

Contents

A few of the authors. Back row, left to right: Ole Arve Misund, Sylvi Inez Liljegren, Stig Lægdene, Tora Hultgreen. Front row, left to right: Marit Anne Hauan, Kjartan Fløgstad, Leif Magne Helgesen, Kim Holmén. (Photo: Jan Sivert Hauglid)

Preface

Welcome, reader! We hope these pages will give you new knowledge about climate, while inviting self-reflection and ethical thinking. Our wish is to inspire you and challenge you to go a few steps farther. Climate ethics is not just passive reading: it is acting purposefully to build a better future. We turn the spotlight on what is currently happening, what responsibilities rest upon us, and we also try to illuminate what we cannot see, but discern as possible scenarios.

This book has several authors, and each will highlight a different topic. Readers may find this miscellany of viewpoints and themes bewildering. Some readers will perhaps be provoked by the absence of certain voices and opinions. As editors, we do not claim to cover the entire field. We are more interested in sparking a dialogue between groups, reflection, and information transfer. Thus the book's authors represent a range of professions within academia, management, the media, natural sciences, the church, and museums. The editors believe in dialogue and cooperation. The climate crisis challenges us to work together across disciplinary, professional and national boundaries.

A book quickly becomes outdated. The facts you find in this book will change over the next few years. Research will provide new answers. Numbers we present here will be adjusted up or down. Dates we mention will soon appear to belong to a bygone era. Developments move fast. Nonetheless we hope readers will allow themselves to be swept along and realize that each of us can influence how the world responds. It *is* possible to contribute. It *is* possible to help create a sustainable future.

Above all, we hope this book will stimulate readers to reflect on ethical issues. Ethical reflection should be incorporated as a fundamental thought process in all fields. The things we do have consequences. The things we choose not to do also have consequences. Our time poses momentous ethical dilemmas. To comprehend these dilemmas, we all need a good grasp of current knowledge. Only then can we make ethical choices. Respect for life – both today and in the future – will be decisive in determining whether our choices are good or not.

We hope climate ethics will be a wellspring of debate and discussion, thus helping build communality that lasts into the future.

We see many of the obstacles that emerge as the ice in the Arctic melts away. Suddenly rocks and islands appear – hazards that have never been on any map. It is vital that we avoid running aground, though the obstacles seem insurmountable. We need a compass that will guide us safely through a sea of troubles. We need to spread optimism about the future. We do that best by charting a course that will minimize negative climate developments.

We thank all the authors and photographers. We thank Eva Therese Jenssen, UNIS, who was in charge of the illustrations. We thank everyone who contributed in other ways. We all believe in a future; but there are challenges we must tackle if we wish to put the most pessimistic prophets to shame.

The editors

Leif Magne Helgesen, Kim Holmén and Ole Arve Misund

Foreword

by Jens Stoltenberg, former Prime Minister of Norway, UN Special Envoy on Climate Change

There are few places on earth where the seriousness of climate change can be seen as clearly as in Svalbard.

At the top of Zeppelin mountain in Ny-Ålesund lies one of the most important observatories for monitoring atmospheric CO_2 concentrations. In 2013, the observatory in Ny-Ålesund was among the first to measure CO_2 levels that exceeded 400 ppm (parts per million) for entire months. In April 2014, all the monitoring stations in the northern hemisphere confirmed that CO_2 concentrations are now above 400 ppm. Levels must stay under 450 ppm if we are to have any chance of avoiding a global temperature increase of two degrees Celsius.

Few places on earth show the impact of climate change as clearly as Svalbard. For years and years the fjords lay frozen in the winter. They don't anymore. In recent winters, Isfjorden and other fjords have had open water all winter. A few years ago, researchers at the University Centre in Svalbard (UNIS) discovered mussels on the coast of Svalbard, and in the autumn of 2013, people caught mackerel in Isfjorden. I have been to Svalbard many times through the years, and have seen these changes with my own eyes.

But there are also signs that give cause for optimism. More and more countries are taking the climate issue seriously and acting to cut emissions. New technology is emerging that enables both economic growth and low emissions.

I belong to a tradition that believes in the usefulness of politics. I think it is possible to solve even the most complex problems. Climate change is one of the most complex issues we have ever faced –politically, technologically, and ethically.

To solve a problem, we need to understand what the problem is, and we need to reflect about which strategies are good ways to solve it. The objective of *The Ice is Melting – Ethics in the Arctic* is to help meet both those needs.

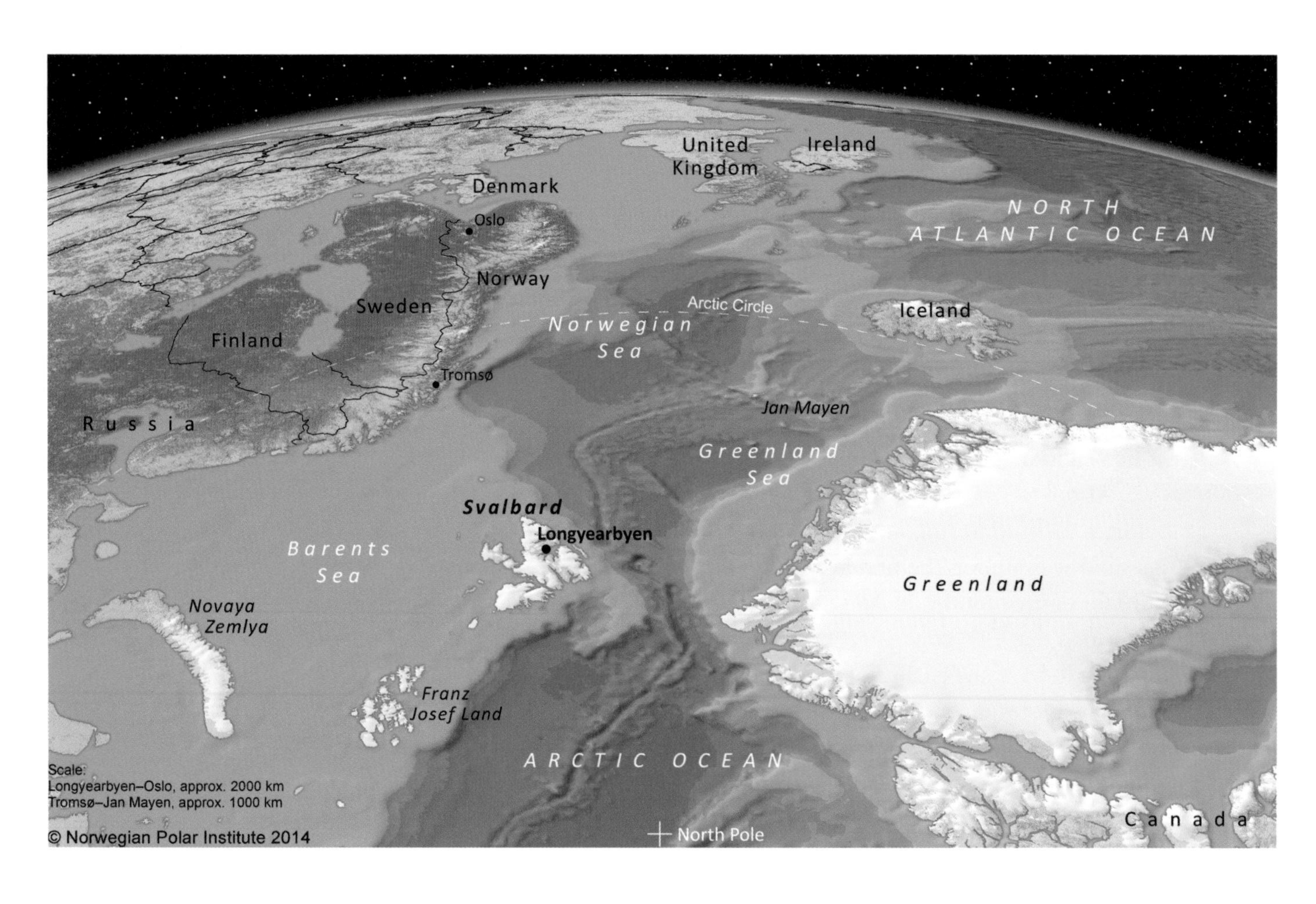

The European Arctic viewed from the North Pole (Map: Norwegian Polar Institute)

The Arctic (Map: Norwegian Polar Institute) →

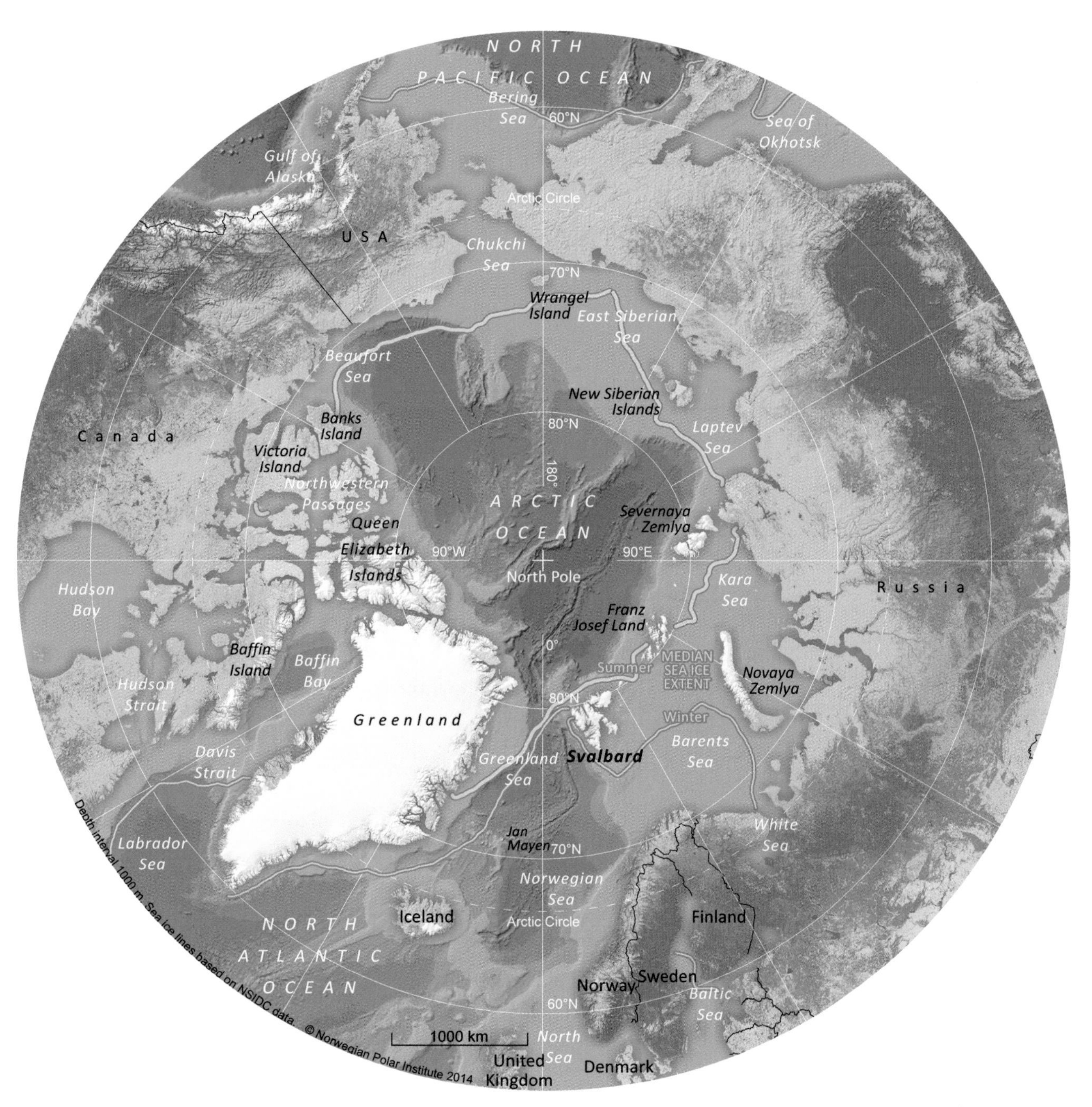
NORTH PACIFIC OCEAN
Bering Sea
60°N
Sea of Okhotsk
Gulf of Alaska
Arctic Circle
USA
Chukchi Sea
70°N
Wrangel Island
East Siberian Sea
Beaufort Sea
New Siberian Islands
Banks Island
80°N
Laptev Sea
Canada
Victoria Island
Northwestern Passages
180°
ARCTIC OCEAN
Severnaya Zemlya
Queen Elizabeth Islands
90°W
90°E
North Pole
Hudson Bay
Kara Sea
Russia
Franz Josef Land
Baffin Island
Baffin Bay
0°
Summer
MEDIAN SEA ICE EXTENT
Novaya Zemlya
Hudson Strait
80°N
Greenland
Winter
Barents Sea
Davis Strait
Greenland Sea
Svalbard
Depth interval 1000 m. Sea ice lines based on NSIDC data.
Labrador Sea
Jan Mayen
70°N
White Sea
Norwegian Sea
Iceland
Arctic Circle
Finland
NORTH ATLANTIC OCEAN
Norway
Sweden
Baltic Sea
60°N
© Norwegian Polar Institute 2014
1000 km
North Sea
United Kingdom
Denmark

Climate, nature and Svalbard

by Kim Holmén, International Director, Norwegian Polar Institute

Svalbard is beautiful, serene, and adventurous. Svalbard is fjords, glaciers, braided streams, permafrost, polar bears, whales, walruses, seals, birds, and flowers: a whole suite of unique and extraordinary features, phenomena and life-forms of the Arctic. Svalbard is one of the last relatively untouched wilderness areas of Europe. Svalbard is an archipelago with a long history of hunting, whaling, and a century of mining. Svalbard is a place that leaves few untouched, a place that many visit, and that some make their home. There are no indigenous people in Svalbard; prior to the mining period human residence was transient, and to a large extent it also remains so. The Svalbard Treaty stipulates that all subjects of co-signing nations have equal access to the archipelago; this gives rise to an international and multicultural group of visitors and residents. In this diversity and magnificence, Svalbard is a place that kindles dialogues about values, ethics and dilemmas as society, commerce, and nature change. We shall in this chapter focus on the changes in nature; other chapters in this book will discuss other aspects of ongoing change in Svalbard.

Arctic climate changes first, most and fastest. This statement is frequently repeated in the debate about human-induced climate change and planetary warming. We study present and future climate change with climate models. These models indicate enhanced change in the Arctic. We study ongoing climate change with measurements and by comparison with historic records; these analyses show that the Arctic is warming about twice as fast as the rest of the world. The consistency between the theory (as depicted in the models) and the observations is compelling. The Arctic is a region of paramount importance for climate research and for enhancing our understanding of climate. The Arctic is also a region where there are clearly visible effects of climate change both in the landscape, in ecosystems and in human lives. Svalbard is one of the areas in this fast changing Arctic where the greatest changes are expected, and seen. We shall expose some of these changes in and around Svalbard: what we see happening and what we may expect in the future.

Climate and climate zones

Climate is a long term average of meteorological variables like humidity, temperature, pressure, wind and others. Weather is the instantaneous distribution of these variables in a region. Climate change is thus a long term shift in one or more of the meteorological variables. By international convention, climate is studied by averaging 30 years of weather. The 30-year convention is chosen based on considerations of known weather variability such that the average catches stable features of the climate. Society is increasingly requesting statements about climatic shifts on decadal or even shorter time-scales. These societal needs create dilemmas for the scientists, since we know that short term variations always have existed. Unless we can determine causality between a particular weather event and some perturbation (which we thus far seldom can on a one-to-one basis) the short term trends need to be considered with caution before placing the blame on human-induced climate change – or attributing them to natural variation for that matter.

One of the classic methods of defining climate zones is the Köppen scheme, which is based on the concept that the native vegetation distribution is the best (long-term) expression of climate. The boundaries in the climate classifications are based on combinations of the measurable quantities temperature and precipitation. Climate zones are distributed around the world in various bands and regions. Historic and future climate changes alter temperature and precipitation distributions. Vegetation and ecosystems have had an ability to redistribute during or following previous climate change since we know that most (if not all) species and ecosystems have existed much longer than the pre-industrial climatic distribution prevailed. Present human-induced changes are rapid and profound; to what extent native vegetation, ecosystems and farming practices

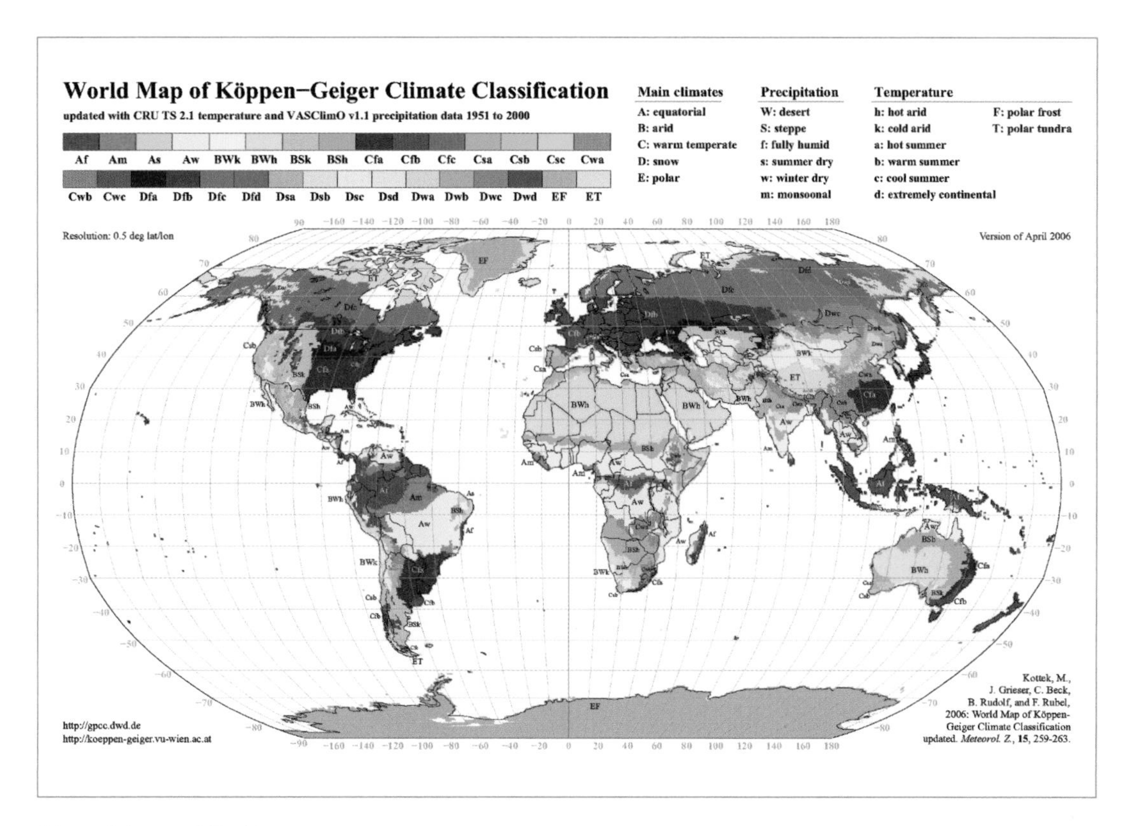

Map of the world's climate zones

can redistribute and/or withstand these ongoing changes is disturbingly unclear. For the polar regions, and the Arctic in particular, the fact that they contain a rare and distinct nature with unique species, ecosystems and physical geographic features is a further consideration. Regardless of their resilience and ability to redistribute, high arctic species will have nowhere to move if the polar climate zone (e.g. ET in the figure) is quenched. As the earth warms further, these entities associated with the polar climate will increasingly lose habitat, until they die out. Not only is the Arctic a region where we expect and see rapid climate change: it is also a region where the impacts of change rapidly approach what may be incurable.

Studying climate change

When studying the effects of humankind's alteration of the atmosphere, and particularly the effects on the earth's future climate, we mainly use models. Today's climate models are computer-based with mathematical descriptions of various processes we have observed in nature, united in an internally consistent numeric package. A model is, by definition, a simplification of reality. Even a "good" model will differ in some aspects from reality since it is a simplification.

Winds are driven by energy. Most of the energy comes from the sun. The earth's energy balance is balanced over long time scales and for the entire earth. However, there are regional and local imbalances – for example between day and night, summer and winter – when energy (heat) is stored in soil, water, or air (or is lost from storage). A permanent and profoundly important feature of the earth is the large influx of sunlight in equatorial regions versus the small influx of sunlight in polar regions. This leads to a positive energy balance in low latitudes and a negative energy balance in high latitudes, as the outgoing radiation from earth (infrared radiation) does not have as large differences between equator and pole. These imbalances and the uneven distribution of ocean and land give rise to temperature gradients that drive the winds. Earth's rotation, the humidity of the air, evaporation and precipitation, friction against the surface will all modify the winds in various ways, and the complexity of these interactions is what makes climate models so difficult to construct.

Models have many weaknesses and uncertainties because some of the simplifications are large, because our understanding or even our knowledge of processes in the real world is incomplete, and because our computers have numeric limitations. The best climate models today are based on the best knowledge we can muster, but they still have weaknesses that are easily exposed. Most of the weaknesses are well known to the scientists building the models, but cannot be avoided with today's tools and knowledge. Identifying unknown flaws and weaknesses in climate models is a positive contribution. Destructive critics dismiss the models by pointing out (often well known) weaknesses, but they fail to suggest a means of enhancing the models. The constructive scientists' obligations are to describe and quantify the uncertainties created by the various inherent weaknesses. The Intergovernmental Panel on Climate Change (IPCC) recently presented its fifth assessment report constructively and prudently.

Climate models (as presented by IPCC) point toward the earth warming from human-induced increases of greenhouse gases in the atmosphere. These results cannot be dismissed despite the uncertainties of the models. Stating that nothing will change because of the alterations we have imposed has little scientific foundation; it is at best wishful thinking, and at worst, an argument of convenience or deliberate disinformation.

The IPCC models single out the Arctic as a key area for earth's climate. The arctic climate is sensitive to changes; the arctic climate will see dramatic changes, and the models also show that changes in the Arctic will have an influence on climate far removed from the Arctic itself. Climate change in the Arctic is a question of global significance.

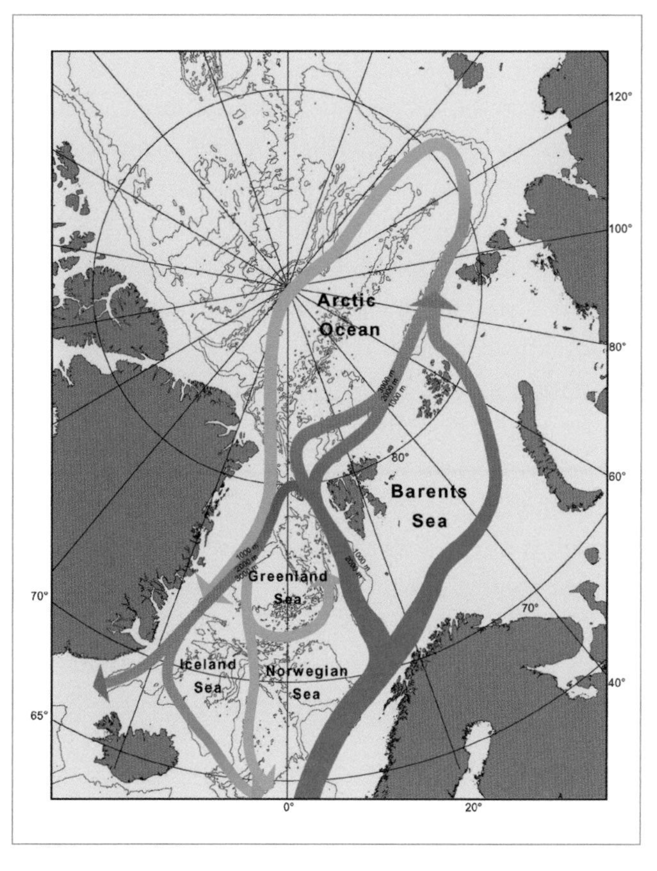

Cold and warm ocean currents in the Arctic (Illustration: Audun Igesund / Norwegian Polar Institute)

Svalbard and climate change

Svalbard is in the middle of important gateways to the Arctic with respect to climate. The figure above shows some important aspects of Svalbard's surroundings. West of Svalbard is the only deep-water channel into the Arctic. The two branches of warm (red) ocean currents feeding into the Arctic pass around Svalbard. These extensions of the Gulf Stream are important sources of heat in the Arctic Ocean. The cold (blue) currents leaving the Arctic are mainly concentrated into the East Greenland current. That the major ocean exchanges between the Arctic and the world oceans occur in the Fram Strait and waters adjacent to Svalbard, make the Svalbard region a highly dynamic area in the Arctic. Surface ocean currents (like the Gulf Stream) are essentially driven by wind; the ocean currents as we see them are manifestations of prevailing winds. The Svalbard region is also a place with proportionally large atmospheric exchange between lower latitudes and the Arctic. Along with its surrounding seas, Svalbard is thus a key location for observing and understanding arctic climate change. This is true both with respect to detecting and determining change, and with respect to identifying and elucidating the processes responsible. Svalbard is placed between the temperate regions and polar regions in a very dynamic area. Svalbard is therefore an archipelago with strong climate gradients and a region with large natural variability in weather. Arguably, the strong gradients and variability can make interpreting changes difficult (i.e. detecting a long term trend against a backdrop of intense variation), but this is also the context in which the dynamics and climate are likely to change the most. However, Svalbard also has a long history of scientific research and is one of the regions of the Arctic where we have the largest amounts of data. All of this combined provides strong arguments for making concerted efforts to study climate and environment in Svalbard.

Svalbard climate

The annual average temperature in Longyearbyen is –4°C. The highest measured temperature in Svalbard is 21.3°C, and the lowest is –46.3°C. This makes it much warmer than any other place at the same latitude in the Arctic. Winds are most frequent and strongest in winter, while fog is more frequent in summer.

The fjords and sea areas north and east of Svalbard are covered with ice for 8–9 months of the year, while the fjords on the west side of Spitsbergen are increasingly ice-free during the winter. Most precipitation that falls on Svalbard comes with easterly winds from the Barents Sea, and the southeast coast of Spitsbergen receives more precipitation than Longyearbyen and Ny-Ålesund.

The four left panels in the figure below show the average temperature during the seasons as based on the 1961–2000 measurements (see NorACIA report for details). We see pronounced temperature gradients across Svalbard, in particular during fall, winter and spring. The four right panels show model results from the same report with calculated temperature changes (in degrees C) when comparing the average for the model period 1981–2010 with 2021–2050. We see that the largest changes are east of Svalbard and consequently the gradients that are so pronounced now will become weaker (the east warms more than the west). We also see a temperature rise everywhere and in all seasons in these climate simulations.

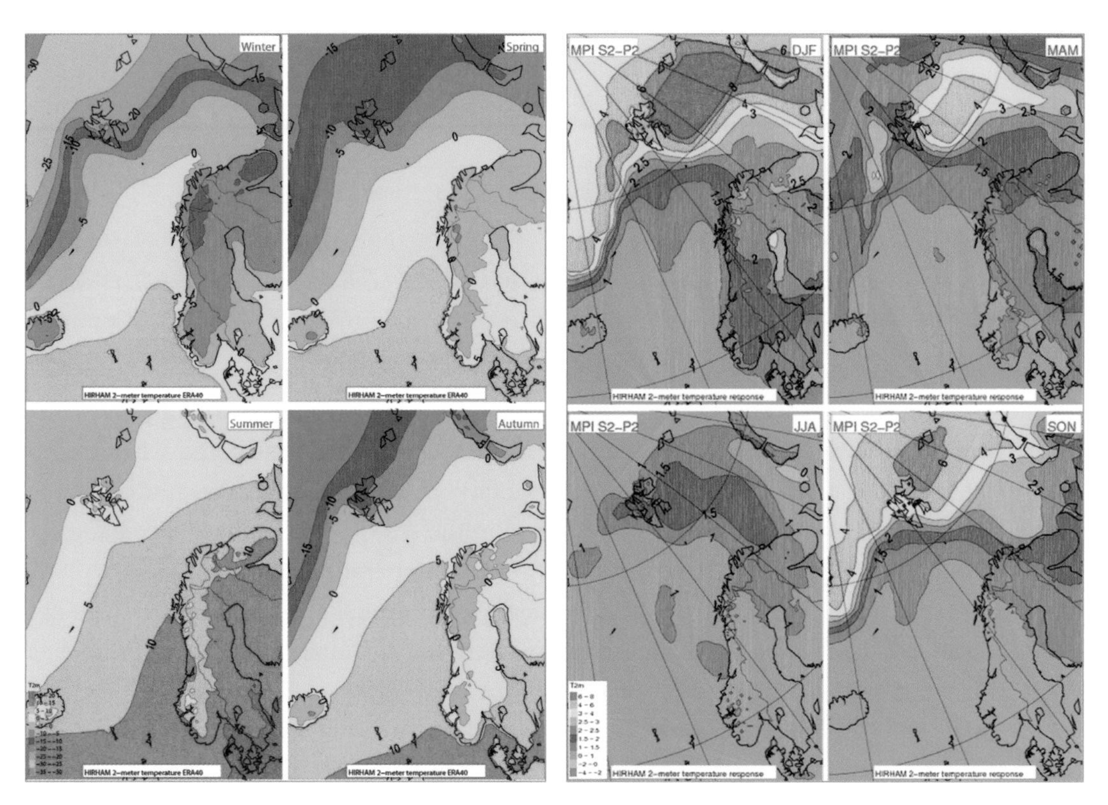

Left panels: Average temperatures in each season based on measurements 1961–2000. Right panels: Model results showing differences in average temperature between model projections for the periods 1981–2010 and 2012–2050 (Illustration: Audun Igesund / Norwegian Polar Institute)

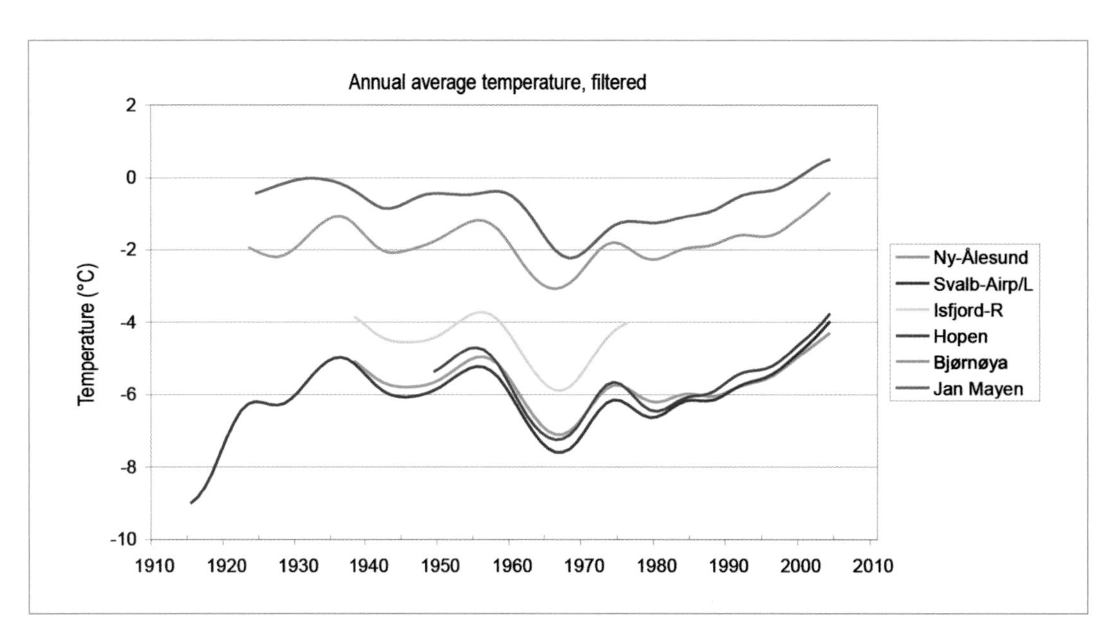

Annual average temperatures at locations in Svalbard 1911–2007 (Illustration: Audun Igesund / Norwegian Polar Institute)

These simulations show calculated future change under a number of assumptions. The uncertainties in the models, and the assumptions made, must be considered when one is looking at maps like these. These are model simulations, and scientists will not call them "predictions" as one does with weather forecasts. But the Arctic and Svalbard *are* changing, and increasingly so. The figure above shows the annual temperature averages for 1911–2007 (also from the NorACIA report). The development of the annual average temperature since 1980 is clear, as is the increased warming in the latest years. Detailed studies of these data show an increase in temperature in all seasons of the year.

Climate and climate change involve not only temperature, but all other weather parameters as well. Precipitation can change both in amount and form (rain or snow). Glaciers are built by accumulation of snow, but there are few things that erode glaciers more than rain. The same simulations that were shown above for temperature also include precipitation. They show increased amounts of precipitation for Svalbard, with the least change in summer, but large increases (60%) on the east side in winter. The warmest areas (west coast at low altitude) are likely to see increased rain and melting of glaciers, whereas high altitudes and the eastern side (which despite having the largest increase in temperatures will still be below freezing temperatures for much of the year) will see increased accumulation of snow. These are indeed patterns the glaciologists are capturing already today in Svalbard.

The landscape is changing, and rapidly so. We see this clearly in the collage below taken from the ridge of Zeppelin mountain just south of Ny-Ålesund on the northwest coast of Svalbard. We must be careful when interpreting pictures like these because of the many factors that control a glacier's behavior. Temperature, precipitation and the fact that glaciers have a long response time to climatic changes all play a role here. Many glaciers in Svalbard

View from Zeppelin mountain in 1922, 1939, 2002, and 2010 (All photos: Norwegian Polar Institute)

are what are called "surging glaciers" that have a dynamic that make their fronts surge forward (by as much as several kilometers) at irregular intervals (of order decades to centuries). Between the surges the fronts slowly recede until the next surge occurs. For the casual observer, the glacier landscape should thus be in continual change. But the collage shows at least six independent glacier systems that would not be expected to surge and recede in unison if their internal dynamics were the sole explanation of the changes observed. Some of the glaciers (in particular the ones to the right in each panel) are known to be non-surging. We can conclude that the collage does indeed show a large scale change of the governing forces for the glaciers in this region. This change has been ongoing for a long time. Some of the changes happened long before the largest changes of greenhouse gases in the atmosphere that humankind has made (the lion's share of the increases have occurred after the Second World War), but the glaciologists are seeing an accelerated decrease in glacier mass during the latest decade. Some of the changes are due to natural variations in the system (a likely explanation of the changes between 1922 and 1939), but we are increasingly certain that the latest accelerated decreases in glacier mass are related to human-induced climatic change.

As the Arctic is warming we see a pronounced change in sea ice. The area and thickness of ice has decreased steadily in the entire Arctic Basin during the past 40 years. There are increasingly clear indications that this decrease has accelerated during the latest decade. Around Svalbard this trend has manifested itself as ice-free fjords throughout the winter on the west coast during the past several years. The polar bear is probably the most iconic species associated with sea ice. But it is only a symbol for an entire unique ecosystem that we find around Svalbard. The figure to the right depicts some of the exceptional features and organisms that are associated with the ice and margins of the sea ice. The polar bear roams on the ice, and anyone who has the privilege of meeting this arctic resident in its natural environment is struck with awe. There are many other residents of the Arctic that deserve to be at least as awe-inspiring as the bear, but are perhaps less visible, less spoken about. The plankton in the sea have special adaptations; the polar cod fry live within the sea ice for protection; the beluga, narwhal and bowhead whales all have special adaptations for swimming in ice-infested waters; the ringed seal has an ability to maintain breathing holes in ice. Then there is the walrus, and all the birds of the high arctic. Snow buntings, little auks, Brünnich's guillemots, black footed kittiwakes and many others all have extraordinary specialties that

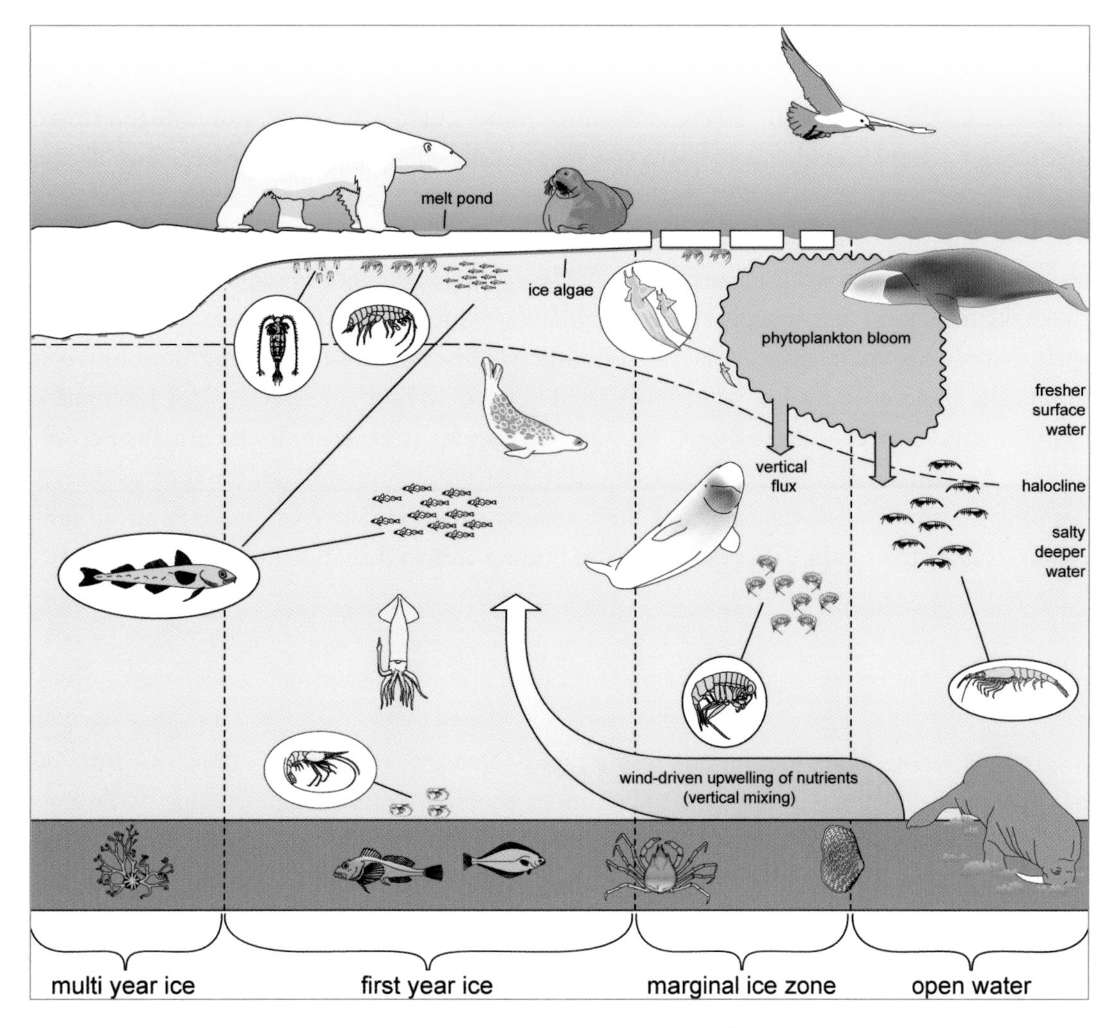

The ecology of the ice edge is diverse and complex (Illustration: Audun Igesund / Norwegian Polar Institute)

allow them to exist in the Arctic. But as the ice disappears they lose their habitat and they will quickly lose out in competition with other species that are better adapted to open waters and warmer land. There are already many examples of species moving in as the climate becomes more favorable to them.

A magnificent species that is particularly vulnerable to these changes is the ivory gull. This completely white, exquisite bird is extremely sensitive to competition and disturbance at its nesting sites. We find it nesting in the most remote places in the interior of the Svalbard archipelago. There is evidence that this species is under pressure; in Canada the population has decreased by as much as 80% during the past decades. The ivory gull is specialized on a diet that can only be supported at the ice edge; as the ice edge moves away from its remote and secluded nesting locations on Svalbard, the distances the bird is required to travel will make securing food increasingly strenuous. A favorite food for the ivory gull is blubber from seals; ivory gulls are frequently seen scavenging on the carcasses of seals captured by polar bears. The preference for blubber makes these sensational birds a species with high levels of pollutants in their bodies as well. They are thus under stress from both climate and pollutants. As climate allows other species to reach the remote nunataks where they nest, they are under strain from disturbances at their nesting sites too. The ivory gull is a more pronounced loser than even the polar bear from human-induced change in the environment.

Svalbard and the future

Svalbard is seeing and will continue to see rapid climate change with significant effects on nature and ecosystems. Svalbard is today under pressure from climate change, pollutants and other human-induced changes. But Svalbard is still magnificent; it remains very meaningful indeed to work for relieving and avoiding further pressures. The Svalbard Treaty provides everyone equal access to the archipelago. In its continued splendor, Svalbard should inspire *all* to meditate, discuss and contribute to ways of developing human prosperity and well-being in harmony with nature and climate. Svalbard is a place to foster admiration of nature, research, knowledge, friendship, and peace.

Elephants on ice

– a fable about the ivory gull

by Leif Magne Helgesen, Pastor of Svalbard

How do you do. I am *Pagophila eburnea*. Such a nice name. It matches who I am. Few people understand my language. But then people are such strange beings. They only understand their own kind, and even that they sometimes find problematic. When I speak, they only hear "crah", as if I am some kind of crow. Crows don't even belong to my family. Where I live, there have never been any crows. It's too cold for them here. When I speak, I say "krieeeh". That isn't hard to understand.

A beauty has many names. So do I. Some call me a *sea chick*, but I am no chicken. Maybe you are beginning to guess who I am. To figure it out, you need to know how I look and where I live. So, who am I? If you care – and I can't be certain that you do – maybe you should try to see more. Then maybe you'll discover I have something important to tell you.

If you ask *me*, I think I look like a dove of peace. An arctic dove of peace, that's what I am. It would be more correct to call me an *arctic gull* of peace. Why does the dove always get to symbolize peace? I dress in purest white, white as our hope for the future. With no black clouds. When I fly swiftly over the ice, I am a symbol of peace and possibilities.

But making peace is hungry work. Without food – good food – even the strongest creature is in trouble. Peace is possible, but without food, it isn't possible. And water. Water is important. Some need water on tap. I need frozen water. Ice on the sea is my dining table. I know the drift ice. Without it, I will have trouble finding food. When the drift ice melts into the sea, my family suffers. It's that simple. And that complicated.

There's a lot you don't know about me. Even those who know something about me don't know very much. The polar bear is more familiar than me. That's because he is big and powerful, like a king. He is frightening and at the same time cuddly, almost like a teddy bear.

Both the white bear and I depend on the ice. I follow the polar bear and eat what he leaves behind. There is enough for both of us, but with the ice gone, we're in trouble. Others will also lose their dining table and may go under. We are all links in a long, long chain of creatures great and small. It's like a chain of Christmas lights. If one bulb blows, the entire chain goes dark.

When I eat what the polar bear leaves behind, the food has traveled a long way to end up on my icy table. A small fish was eaten by a big fish, which was eaten by a seal, which was eaten by a polar bear – and me. And there are even more links in the chain. The food came from far in the south. This is nothing new. My ancestors also ate food that came from afar.

Nesting ivory gulls (Pagophila eburnea) *(Photo: Kim Holmén)*

The ivory gull has become a rare sight in Svalbard (Photo: Kim Holmén)

The problem is that humans are now dumping so many strange things that don't belong in nature. What doesn't belong in nature turns into poison, moves northward, and ends up on my plate. It's causing me problems. The eggs I lay have thinner shells now. I don't lay as many eggs as other gulls. That makes me especially vulnerable. When the shells of my eggs get thinner, fewer gull chicks survive and grow up.

Our family has shrunk to a tiny group. If this continues, we will die out. Then you will only see me in pictures, the way you have pictures of dinosaurs.

In English they call me "ivory gull". In Norway, where I often nest, *gull* means gold. Most everyone understands gold. The quest for gold has driven people on expeditions since time immemorial. My other name – ivory – shows where I live. I feel most at home in white landscapes. I myself am clad in white, with black legs. I live far in the north, farther north than any of my fellow creatures. You might say I am on top of the world.

If you ever see me fly, count yourself lucky. Not everyone has that experience. There are only a few of us. I don't know how many, since we are spread over such a wide area. We live in the north, preferably far from humans.

The white gull in the white landscape. White as ivory. Thus the name *ivory*. Humans hunt for ivory. That happens far outside my own world. There are no elephants on the ice. What happens far away doesn't concern me. At least that's how humans think. Sadly I know better.

Can you imagine an elephant on the ice? That would be a funny sight to see. With such big feet she would certainly be able to walk, but what if she slipped and her legs splayed out like an acrobat's? Her trunk wouldn't be much help. With all her legs straight out it can only end one way. There would be a big crash. The ice would probably crack and the elephant would end up in the cold, wet water.

The ice is often covered with snow. That's why it's white. If the elephant looked back as she walked on the ice, she would see huge tracks in the snow. That's how I picture humans. They leave huge tracks. But it isn't often they look back and discover they made those tracks themselves.

Ethics in the Arctic

by Stig Lægdene, Principal, Northern Norway Educational Centre of Practical Theology and Leif Magne Helgesen, Pastor of Svalbard

Climate ethics

Is it possible to put the words *climate* and *ethics* side by side, or do these concepts belong to two different worlds?

Climate indicates something tangible, something we can see and feel. It speaks of weather and the physical reality around us.

Ethics is about theories as guidelines for how we should live. Ethics denotes a theoretical discourse about the relationship between right and wrong, good and evil. While ethics is a theory, morals guide our actions. We often use the terms *ethical* and *moral* interchangeably. That says something about how theory and practice should be intertwined.

It would be pointless to talk about climate ethics if we believe that humankind has played no part in what is happening around us. But if the weather and natural cycles change because of human activities – which currently available research shows to be very likely – climate ethics is an important concept. The concept deals with how we humans should live to avoid having a negative impact on how nature and the climate develop. This makes climate ethics one of the most crucial concepts of our time.

How people view and relate to climate has major repercussions for life on earth both today and in the future. Our climate ethics will affect both our own lives and those of generations to come. That puts it in perspective. Our time has intimate links with ages to come.

Longyearbyen basks in the evening sun in April (Photo: Diana Snibsøer)

Our steadily growing ability to wreak havoc – both in our own time and for future generations – is frightening. The actions we take, and those we neglect to take, may have dramatic consequences for posterity. Although this is frightening, it also offers many positive opportunities. We can influence the earth in a positive direction, so it can continue to sustain life. We can choose a pessimistic view of the future, or we can focus on the possibilities that spring from making the right choices.

Life itself is threatened. Over the past few years, scientists have noted dramatic changes in the climate. Warning bells are ringing. The climate issue is today's most challenging ethical dilemma, alongside the chasm between rich and poor – and poverty and climate change pull in the same direction. Both poverty and climate-related adversity will most likely cause mass migration and conflicts. US Secretary of State John Kerry, speaking in the Indonesian capital of Jakarta in February 2014, listed human-induced global warming among the "global threats" and ranked it "right up there" with weapons of mass destruction.[1]

Water, food and energy are critical resources, and they always will be. Future conflicts will center mainly on these basic human needs.

Our short frame of reference is a hindrance in discussions of climate and environmental issues. A couple of cold winters suffice to make us dismiss the climate crisis. In Svalbard, many breathe a sigh of relief when ice forms at the inner end of the fjord; it is so tempting to believe all is well with the world.

Few societal and moral issues cry out more insistently for a long-term view than climate change. The changes we currently see in the climate are strongly linked to human behavior. Clearly we must also take natural variations into account when we try to understand what is happening and what lies in store. Nonetheless, the part human activities play in climate change means that our choices will have an impact for life on earth for decades, centuries and millennia to come. That gives perspective.

We can also be deceived by narrow-minded geographical thinking. Geography as many understand it – boundaries between countries and states – is a simplification. Climate issues cannot be confined within the limits of any individual region. What happens in one place will have repercussions elsewhere. When ice melts in the Arctic this has dramatic effects both locally and globally. Local events must be examined in a broader context. What happens in the atmosphere will influence the oceans and vast areas of the earth. Everything is interrelated. Climate ethics means solidarity that transcends geographical borders and generations.

The Arctic

The Arctic serves as a barometer of climate change on earth. If the world's average temperature increases by two degrees Celsius, the temperature will increase far more in the Arctic. We see these changes far earlier in the north. A report tells us "Arctic average temperature has risen at almost twice the rate seen in the rest of the world in the past few decades."[2] In the Arctic, we can peer into the future. We can see indications of how animals and plants will adapt to the changes – or succumb to them. This is a vulnerable habitat. The Arctic is at the outer limit of where life can be maintained. We are at the end of the road. In other places, some of the plants and animals can migrate when climate changes, but for those specialized to survive in the coldest climate, there is no way out when it gets warmer. Small changes can have major consequences. This makes it imperative for humanity to do research and keep close watch over climate developments in the Arctic.

What will happen if temperatures steadily rise? How will the permafrost react? What effects will we see on coastal erosion? What will become of the glaciers? What will happen to the vegetation and life forms in the seas? Researchers have some of the answers, but are not able to predict our course into the future. We still do not fully understand many of the natural systems that function in our own time. Even more unanswered questions arise when we try to look into the future.

What happens in nature affects humans too. Warming will change work and leisure activities of people who live in the Arctic. This is particularly true for those who rely on traditional ways of life. The Arctic is home to about four million people. Of these, 320,000 belong to indigenous peoples,[3] and most of them live by hunting and fishing. They are already under severe pressure because of climate change. Sofie Petersen, Bishop of Greenland, who is herself an Inuit, gave a speech in Svalbard, where she described the observations of a hunter's wife from Qaanaaq/Thule:

> Here in northernmost Greenland, where we live in a hunting community, we have noticed climate change in recent years, because we have been forced to change our way of life. When the ice grew rotten, the hunters had to become fishers. The thick ice, where we used to be able to sled great distances to hunt the large animals – that ice has changed. Our old tradition of going out to hunt with dogsleds that do not pollute – that is in danger now. Our people's food, our people's clothing, our people's livelihood – all these are threatened.[4]

More and more countries and commercial enterprises are eyeing the opportunities the Arctic may offer. Economic interest is on the rise. Many corporations are keeping a close watch on developments. Current ethical issues in the Arctic include the dilemmas related to shipping, fisheries, the petroleum industry, mining for coal and minerals, and carbon capture. The choices we make in the Arctic are local choices with global implications. The consequences of increasing pollution may have a dramatic impact on life here in the north, and for the planet as a whole. This makes ethics imperative. We must ask the difficult questions and lay out our options. It is vital to provide opportunities to choose.

Four million people live in the Arctic. Longyearbyen is home to about 2000 of them (Photo: Diana Snibsøer)

Responsibility

It is in the nature of ethics that it should provoke and help us ask the right questions. Ethical deliberations will challenge us both as individuals and as a society.

Climate ethics will demand many important political decisions. This makes it crucial for citizens to exert their political power by voting. Climate change is one of the fateful challenges of our time. We should therefore expect the political parties that govern our countries to be deeply committed to climate issues. As voters, we can make our opinions known.

If election campaigns are not focused on climate issues, that is not because the challenges have become any smaller, but because some politicians ignore them. It is unfortunate when elected representatives do not take the climate seriously, but opt for populist solutions.

At the same time, many politicians are dedicated to averting negative climate trends, but need voters who support their view that climate policy is more important than most other issues. Politicians in a democracy rely on public opinion demanding new solutions.

Many politicians despair at how little attention is paid to climate and environmental issues during elections. Yet the politicians in power must bear the heavy responsibility of making correct decisions about the future. Politicians who choose to ignore these issues are doubly culpable if the situation develops in the wrong direction. The question is how we as voters can support politicians when they must make tough decisions. How can we give them the courage and strength to stake out paths leading to goals that will not be reached until long after their term in office? The answer lies in using the ballot box, but also in fostering a *political* climate that favors policy decisions that benefit the *global* climate.

It is problematic that terms of office and election cycles discourage politicians from thinking more than a few years into the future. Short-sighted political goals – when they impinge on climate – will have long-lasting repercussions. Important political decisions must not be impeded by the fact that there is an election every four years. We need politicians who are capable of making the right decisions, even if those decisions come at a cost.

Commercial interests in the Arctic are a challenge. Nature's equilibrium may be upset if humans perturb the system too much or too long. Each intrusion, each extraction of natural resources, calls for accountability. If we harvest too much fish from the Barents Sea, crucial stocks may be threatened. But the Barents Sea fisheries also provide a good example of how management and bilateral agreements can be successful in the Arctic (see the chapter by Ole Arve Misund, page 213). A far greater challenge is the growing interest in oil and gas extraction, mi-neral recovery, and shipping. These enterprises require international regulations and agreements. But we also need an open discussion about fossil fuels. Should we exploit them less? Should we ban exploitation altogether? How can we deflect a trend that increases harmful greenhouse gases in the atmosphere and will affect life for many generations?

Companies operating in the world's most vulnerable regions carry the heavy responsibility of ensuring that they leave no permanent tracks. On top of the obligation to tread softly here in the Arctic, the companies have global responsibility. Ethical considerations should be part of the groundwork before an activity is initiated in the Arctic. The question is whether commercial enterprises value adherence to ethical principles when making critical decisions, or if economic considerations hold trumps. All activities in the Arctic are associated with ethical responsibilities.

Climate change also challenges us in our day-to-day lives. Every day we make choices, large and small. Should we walk to work or drive? Should we use plastic bags from the store or take along reusable ones? Do we turn off the lights when we leave a room, or are our homes and offices constantly illuminated? Everyday life presents a series of choices with ethical dimensions.

We need a responsibility ethic that applies to individuals, commercial interests, and political leaders. We must be accountable for our own actions and those of our generation. The fact that we are aware of the consequences increases our responsibility. Affluent nations and people in positions of leadership have greater responsibility than farmers and fishers in poor countries – but the latter are often those who feel the changes most directly. Climate change places special demands on affluent nations. Our consumption of natural resources makes us accountable.

Our responsibility is far greater than we have been willing to admit so far. Climate change calls for action. Climate ethics can become a catchword in political speeches and seminars on ethical business practices. But if the ideas move no farther than theories and pretty words, we will have lost. Ethical thinking challenges us to practical actions.

Climate ethics traces out a blueprint for the lives we must lead if humankind is to survive to future generations. Carpenters know how important blueprints are for constructing a building. Similarly, ethical thinking about the climate is fundamental to the adjustment process we are already going through. But unlike construction work, where only architects and engineers are involved in planning, the climate issue impels us *all* to increase our awareness and participate in tracing the blueprint for our future. We mustn't turn all responsibility over to politicians and government agencies, even though they bear heavy responsibility – of which they will hopefully prove themselves worthy.

Many of the ethical choices we make on a daily basis have immediate effects. That is not always the case when we make choices that will influence the Arctic. Many people live by the motto *Carpe diem* – seize the day. If this prompts us to live only in the present, without a thought of tomorrow, our descendants will have a problem. That way of thinking is egoistic. A motto like *Carpe diem* must never ignore the community and the future: our actions have consequences for both.

More than any other issue, climate change forces us to act against a backdrop of long-term ethical responsibility, one that acknowledges that the future existence of humanity and of nature itself is the fulcrum on which we must balance our choice of action. It is heartening that we, as rational beings, have an opportunity to make sensible choices in matters that will reach far beyond our own lifetimes.

Trivialization

Short-sightedness impedes choices that are good for the climate. Other obstacles include attempts to trivialize the situation, or simply explain away disagreeable viewpoints.

The essence of research is constantly to seek new knowledge and to revise hypotheses. Research is forever on the move. The truth can be both complex and multifaceted. But that does not mean we can sound the all-clear just because a few scientists have completely different interpretations of what underlies the climate change we already see. Great responsibility lies on the shoulders of researchers and politicians who choose to ignore the danger. The future will tell who was right, but by then it will be difficult – maybe impossible – to compensate for misguided decisions.

Maybe we need a better understanding of what science and research entail. Science rarely deals with absolute proof and inalienable truths: it deals with degrees of probability. Debate and contrary interpretations are central to science. This does not mean that all viewpoints are equally important, or that all results are equally credible. A handful of critical voices tend to attract just as much attention as 2000 scientists who say that human-induced climate change is highly likely – as they have stated through IPCC, the Intergovernmental Panel on Climate Change. The media often exaggerate disagreement, and some political parties would rather adopt populist attitudes than listen to unequivocal recommendations from international experts.

It is tempting to shrug off our emissions – to think they can't be all *that* bad. So we bury our heads in the sand and hope for the best. We forget that it will be a long time before climatologists can verify the calculations they have made based on what science currently knows about the laws of physics. If we trivialize climate change today, and do nothing, the future detrimental effects will be more severe than they would have been if we had taken precautions.

An overwhelming majority agree that the climate is changing. The question is whether the changes reflect nature's own variability, or the impact of human activities. The work of several thousand researchers, compiled by IPCC, shows that it is highly likely both that the climate changes because of natural variations, and that humans influence climate in a negative direction. Against that backdrop, it is unethical to disregard human activities as an important factor in ongoing climate change.

Moralization

Facing the fact that the ice is melting, acknowledging the dire consequences this may have, and yet believing it is possible to act now to benefit the earth's future – this is not highbrow zealotry, as some mockingly claim. This is realism, coupled with a positive outlook on life and human potential. Those who deny all responsibility and reject any possibility of making changes for the better, apparently underestimate human intelligence and resourcefulness.

Moralistic preaching can counteract sound ethical choices. If we elevate a particular lifestyle to the status of "right" and impose it on others, we are moralizing. And moralizing rarely serves as anything but a provocation.

Ethical accountability in issues of climate does not mean we must all conform to a single standard. People have different needs, different aspirations. Rather than prescribing one specific lifestyle to avert the climate crisis, we should encourage ingenuity and diversity, seek inspiration in success stories, emulate role models.

Moralization restricts inventiveness, whereas knowledge and information open up broader perspectives. We should strive for a creative, life-affirming attitude that brings out the best in people rather than suppressing them.

Some lifestyles are inconsistent with good climate ethics. As citizens of one of the world's affluent countries, Norwegians all live in glass houses – but sometimes it is necessary to throw stones. Ongoing developments are too serious for compromise. The challenge is figuring out how to pass on important values in a positive way.

Above all, we must never give up on finding ways to live in harmony with nature, and the choices we make must respect both human dignity and nature's capacity for resilience.

History gives many examples of people who have transformed their present and future – and made the world a better place. Many of these changes began with small steps. Each and every one of us can apply our ideas and climate ethical values. If many people take many small steps, it will add up to great progress. We can all change the future. That is our privilege as human beings. We have possibilities.

Ethics across boundaries

Climate change and development raise many major ethical dilemmas. A central question is how we should relate to commercial development and economic growth. Is it possible to focus on climate and simultaneously expand a business enterprise? Are commercial development and responsibility for the climate on a collision course, or is it possible to unite positive forces and create a sustainable global community?

It is a major challenge that we as consumers often lack alternatives: we have no environmentally friendly choices. If such choices are nonexistent, even the most well-intentioned individual cannot do otherwise than purchase the old products. This gives us no real choice. Without climate-friendly products, we are forced to make unethical purchases. Manufacturers must be innovative and produce goods that give good new alternatives. Politicians must use their power to encourage and support development of new green products.

Unfortunately, most lobbyists appear to be on the side of outdated technology and yesterday's products. Manufacturers are often complacent, trapped in conventional thinking. It is much easier sticking to business as usual than devising something new. Innovation costs time and money. As consumers, we need to act with deliberation and demand that manufacturers and politicians help society move in a direction that benefits climate stability. We must protest against the unavailability of environmentally friendly products.

Are any ideals too precious to be questioned? Can the pristine quality of a region ever exceed the value of all the resources it might have to offer in the short term? Will the climate tolerate further extraction and combustion of fossil fuels? Does the current situation force us to rethink?

Asking questions and pondering dilemmas is the very essence of ethics. We need to challenge each other to ask questions. There is more reason for fear when questions remain unasked, and when development is based solely on short-term economic growth.

To give impetus to fruitful thinking about values, there should be greater emphasis on bringing various communities together to discuss issues and dilemmas. Communities often isolate themselves, forming cliques. This leads to formation of different cultures that do not communicate with each other. Ethical thinking can be a challenge, but it also invites a dialogue.

Natural science and religious thinking have often been on a collision course. Even if we have different standpoints concerning religion and beliefs, we can meet in ethical discussions. The climate issue is so important that we must cooperate and reason together across all boundaries and conflicts. The diversity of world views and political beliefs need not be a threat. The climate issue forces us to confront our fear of people who think differently. If we are to get out of the rut and bring about real change, we need to cooperate.

If we do not bother to cooperate, the threat to the climate grows. Indifference and deliberate trivialization of the climate situation threaten human existence. The issue of climate – more than any other issue – demands that we cooperate over the religious, political and cultural lines that divide us. That is when things start to get really exciting. Dialogue across boundaries and between factions is a creative journey. It is important to set up arenas where it is possible to engage in dialogues that reach beyond the media's need to create confrontations.

Ethics for individuals and communities

Climate ethics is all about how we behave as individuals and as a community. Just as we as individuals have our personal values and attitudes, societies adhere to certain sets of values. These guide our actions and our politics. Climate ethics thus encompasses both individual ethics and collective ethics.

The individual side of climate ethics informs what each of us does – and what we do not do. Small, seemingly trivial actions are important. Major changes start out small. A tiny snowball can become huge – for better or for worse.

Personal ethics and societal ethics are intertwined. If nobody does anything at home for climate and the environment, it is hard to envisage any scope for change in the community. If individuals make ethical choices, but communities do not, the overall effect will be small. Without engaging entire communities, it will not be possible to bring about changes that will significantly affect climate. The challenge is to establish a climate ethic that embraces both individual and societal ethics.

Good climate ethics requires that I myself take action, but to bring about real change, many must strive for the same goal. When groups start moving in harmony, it will open up opportunities for political decisions and binding agreements. But to establish a forward momentum we must all to work together. Then, like ripples on water, we can swell into a wave of change.

This is no easy task. Persuading many people to move in the same direction is a challenge. There will always be some who do not want to participate. In particular, many will refuse to support decisions that affect their wallets. If the changes appear to encroach on what people perceive as their freedom, many will pull back. It all comes down to which values we choose to build our lives on. Ethics is precisely that: choosing one's values.

Another factor is that strong economic forces use populist propaganda. Large national and multinational companies spend substantial sums to present their activities as environmentally sustainable, though this is sometimes no more than an illusion. The hallways of power are full of lobbyists. It can be tricky seeing through this charade and winning against the power of money. It is like David and Goliath. But history tells us who won that fight.

Role models and success stories help create a movement. But they are not enough. Sometimes strong political action is needed to bring about important changes. Price increases or outright bans may become necessary. Other times, it helps to lower fees or grant privileges to those who choose climate-friendly products. Examples are the Norwegian policies aimed to promote the use of electric vehicles: lower annual fees, permission to use bus lanes, and free parking.

There are obvious connections between ethics and politics. Ethics is the edifice of a society's moral code, and politics is its instrument. Yet that is not the entire truth. Politics also feeds back into society's moral code. These two always interact.

Sorting and recycling of trash

Trash sorting is an instance where the individual's ethical actions have consequences for the community. Norwegian municipalities have decided to sort and recycle trash. This entails both investments and savings. Ultimately, these are ethical choices that have implications at community level. As individuals, we can choose to heed the call and sort our trash. We can also choose to ignore it. If we ignore it, we may be fined, and our trash may not be collected. The individual's ethical choices are supported by a societal ethic that says we should sort our trash. If we choose not to, it may be because we don't care, or because we disagree with the municipality's routines and politics. The consequences of one or two individuals not sorting their trash are negligible, but if many start shirking their duty, it will be more costly for everyone.

Regardless of what we do, we are making ethical choices with implications for the climate. Doing nothing is also a choice. When we decide to ignore the consequences and just let the ship sail on, our choices lie behind that decision. We choose to let the ship sail on, come what may, or we take the helm and steer.

Both doing something and doing nothing are active choices. Not sorting your trash can be called unethical, but it is nonetheless a choice. Ethics tells us which choices are good. What is right and what is wrong? What is a morally good act, and what is immoral?

In a deadlocked conflict, it is important not to lose faith in the possibility of peace. Then, even in the darkest situation, we can gaze up at the horizon and focus on the possibility that things will improve. Peace is possible. Even when the future looks bleak and climate projections point the wrong way, we must not lose our faith that we can alter the future.

Many small deeds for a better climate create a movement that will build the future. In this way, trash sorting and recycling contribute toward limiting the amount of trash in the ocean. In addition, handling trash makes us aware that even small deeds help. Our belief in the future starts with the actions we take today.

Anti-smoking laws

Norway's anti-smoking laws[5] are an example of how societal ethics can have consequences for individuals. The law was introduced in 2004. Smoking has many similarities to the issue of climate, and serves as a valuable example. Anti-smoking laws prohibit people in Norway from smoking indoors in public places.[6] The law was based on both health concerns and ethical considerations. It had been proven that smoking was dangerous. Politicians wanted as few citizens as possible smoking or being exposed to passive smoking. Behind the politicians stood the electorate. Among the politicians were individuals who dared to take up the fight because they saw it as an important ethical choice. The potential socioeconomic and health benefits of banning tobacco were so great that some made the effort required to lead the fight for a smoke-free society. In this campaign, they had to expect ridicule and foul play on the part of powerful opponents.

The fight against passive and active smoking was fought against an aggressive tobacco industry. This industry currently directs most of its propaganda toward countries that lack the legal protection consumers now have in more developed countries. Companies like that will always target those who are most susceptible, to maximize profits. Morality weakens in inverse proportion to the amount of money that flows into the bank account.

Through this law, society made smoking indoors a social issue. It is likely that the law contributed toward changing people's opinions and set a new standard for individual ethics. The law made smoking less attractive. An activity some considered an inalienable right – to smoke anytime, anywhere – was suddenly no longer allowed. This was a different attitude, brought about with the new law as the instrument of change.

Nowadays few find it odd that smoking is not permitted indoors. On the contrary: we react if anyone lights a cigarette in the office, or when we enter a living room where there are smokers. If children are present, smoking is inconceivable in most homes. That restaurants are now smoke-free seems like the most obvious thing in the world. These formidable attitude shifts happened over a very short time. The unthinkable is now possible.

These shifts in society's attitude probably mean that fewer individuals choose to smoke. Societal values influence individual values, but also vice versa.

The anti-smoking law did not write itself. Its enactment was preceded by a long ethical debate about smoking and its consequences for individuals and their surroundings. Years of debates and lawsuits paved the way for political decisions. Forty years ago we already knew how dangerous smoking was, but it was politically impossible to implement changes. After many years of awareness-building and debate – also about ethical aspects – suddenly one day it was possible to implement the new law. Nonetheless, the decision to enact strict anti-smoking legislation was controversial and required action from individuals. Courageous politicians had to stand up and support the law.

This example illustrates how community ethics and individual ethics have complemented each other, like instruments in an orchestra. The result is fewer smokers, which most of us perceive as something good. Ethical reflections and moral actions have walked hand in hand.

How can we advance opinion where climate and the environment are concerned? How do we make it equally unacceptable to let the car run idle as it is to smoke indoors? How shall we make taking the bus perfectly normal and a better alternative to driving a car? How do we get to where it seems just as eccentric for a political party not to have an active climate policy as it is to wish to go back to the time before smoking in many public places was banned? How can emission of greenhouse gases become just as unacceptable as smoking in the lunchroom?

Change is possible

It is easy to lose heart. Is it too late to do anything? Will it be any use? What if others don't follow along and I end up a lonesome oddity among my contemporaries? What if people laugh at my lifestyle?

History fortunately has many examples of how people's individual and community efforts have made the world a better place. These are pathways communities have trod before. We have examples of how huge changes for the benefit of humankind have been brought about despite overwhelming opposition. The struggle to abolish slavery engaged both individuals and communities. They were pitted against strong adversaries. Parts of the commercial sector in the 18th and 19th centuries resisted bitterly. It was crucially important that individuals stood up for their beliefs. Some of them gave their lives for the cause they believed in. It was also crucial that society decided to formulate laws that banned slavery.

Many contexts remained where enlightenment needed to spread, and deep-seated attitudes needed to change. Working for important social reforms undoubtedly looked hopeless for long periods of time. But reform came in the end. Examples include women's rights, the struggle for a decent wage and decent housing for workers, acceptance of homosexuality, the fight against racial discrimination in the United States, apartheid in South Africa, and bigoted notions about the Sámi population in Norway. Change is possible, but also demanding. Many examples show that the work is not finished simply because a new law has been enacted. Keeping ethical values in focus will always be an ongoing project.

We can find many cases where individuals and movements believed in an idea and brought it to pass. Their actions and thoughts led to changes that were important for the future. Courageous politicians made difficult choices at times when triumph seemed unlikely. Many people never lived to see their convictions bear fruit, but they made conscious ethical choices for the future. They had insight. They did an ethical analysis and acted according to their moral code.

We who live today reap the benefit of what they sowed. Much that they struggled to achieve, we now take for granted.

In the issue of climate change, it is now our turn to make ethical choices. It is our turn to distinguish between right and wrong. This is the most penetrating question we can expect from our grandchildren and great-grandchildren. They will read history and judge us by what we did – and what we failed to do.

There is a lot of hopelessness in the debate. Some people say that the little they can contribute will make no difference – that it is too late to do anything. It is true that even if we were able to stop all emissions this minute, the climate would continue to change for decades to come, until the atmosphere's composition – already perturbed – returns to some kind of equilibrium. We must adapt to climate change, whether we like it or not. Nonetheless, we need to do what we can to limit the damage and save what we can.

Some people become passive. Hopelessness and apathy are negative forces in the debate about climate. The feeling that it is "already too late" can lure us into living on as before, without sparing a thought for the consequences. We need to inspire each other, encourage each other to take responsibility and do what we can to ensure a better future. It is not too late.

Someone must lead the way. Someone must believe in a change for the better. Together we can lay a foundation for necessary actions. We can support the individuals and movements that stand at the forefront, creating a secure future. It is equally important that we ourselves take responsibility and make ethically sound choices. It is still possible to do something. We can help kindle new hope for the future.

RV Lance *is approaching the edge of the ice (Photo: Sylvi Inez Liljegren)*

Meanwhile, the kitchen is on fire

by Sylvi Inez Liljegren, journalist, Norwegian Broadcasting Company (NRK), and writer

A text message arrived on my cell phone today. “Help the victims of the typhoon in the Philippines.” I agree to donate, mostly from habit, and text “EMERGENCYAID” to some number or another. What’s 200 Norwegian crowns? Enough to help another human being? Enough to put out a fire? Scarcely.

“Good afternoon. You’re listening to the twenty-four-seven news on NRK. Norwegian rescue workers are now on their way to Manila, among them an expert on water purification, because in many areas there is an acute lack of food and drinking water. Soldiers have just shot two suspected looters. Both the army and the police are moving into districts where total chaos reigns.”[7]

I’m drinking coffee at my kitchen table. Here it is warm, dry, and cozy, but before the end of the broadcast, I’ve heard that the stench of corpses has begun to spread in the city of Tacloban, where more than 10,000 people are feared to have perished.

“The full scope of the devastation wrought by supertyphoon Haiyan in the Philippines remains to be seen,” writes the daily paper *Dagbladet* on its website, alongside a photo of little girls in tears.

A ding signals the arrival of a text from SOS Children’s Villages, begging for more support. I write “NEED” on my cell phone and donate another measly 200 crowns.

Washington Post journalist Max Fisher notes that the authorities have no overview of the extent of the damage and have been unable to organize national efforts to assist people who have lost their homes. He stresses how the storm’s devastating power took the country by surprise.

"Why wasn't the Philippines more ready? About 20 tropical cyclones hit the country every year, making it practically a routine. The arrival of typhoon Haiyan was certainly no surprise," writes Fisher. "But the fact that its officials and agencies have not even come together on the death toll, much less a national effort to serve the hundreds of thousands reportedly displaced, highlights just how badly the country was caught off guard by the storm's destruction."[8]

This is about poverty and inadequate logistics. But it's about far more than that. It's about the changes in climate that the world still refuses to take seriously. Not because we are unable, but because we are unwilling. And in this lack of will lies a huge ethical dilemma. Is it permissible to feign ignorance?

A kitchen and a coffee cup

I finish reading the article and pour another cup of coffee, wondering why we are so poorly prepared. And now I don't mean the Philippines, but each and every one of us. Countries around the world. Politicians. Aid workers. Churches. You and I. We have been warned. Researchers have been trying to tell us since the 1980s that the climate will change. Everyone knows about greenhouse gases nowadays. But we don't want to listen. Every time the media tried to let someone raise a warning finger, some critic got a chance to present counter-arguments. This is called "equal time" and is basically a sound ethical principle in journalism; it maintains balance. But for years the message about the approaching climate crisis has drowned in a futile debate about whether climate change was caused by humans or not. And scientists – let's face it – have never been very good at pithy statements. Many don't dare say anything pointed, afraid of looking foolish in the eyes of their colleagues. Afraid of being wrong. But things are changing.

In 2005, the Arctic Monitoring and Assessment Programme presented an extensive report on environmental toxins that attracted much media attention. This report also mentioned climate issues. Still, in my opinion, journalists appeared more concerned about maintaining a balance between the report's proponents and detractors, than about its factual content. And researchers have been reluctant to make any claim they cannot prove with 100% certainty – something no researcher can do. It's in the very nature of scientific research.

We journalists have always focused more on *reporting* events than on *explaining* them. Or warning. Raising a warning finger is not really our job. We must never sermonize. We express concern about today's events and start afresh the next morning. And day follows day, and years go by. Recent research shows that people in Norway are less concerned about climate change now than they were just a few years ago. Those asked to explain why this is say that the entire climate crisis is too "far away". That it doesn't involve us. They say people believe it's all about what will happen in other countries, not at home.

I don't believe that explanation. Severe damage after storms in western and northern Norway – that's not far away. New York, Boston, and Chicago are not far away in our globalized world. Not Thailand or the Philippines either.

Right before our eyes, extreme weather, storms, cyclones, and typhoons sweep in over islands and land. They are called Dagmar in Norway, Katrina in New Orleans. Or the 1970 Bhola cyclone that devastated East Pakistan, now called Bangladesh. This is nothing new. *Wikipedia* tells me the Galveston hurricane of 1900 was worse than Katrina. I don't know how they compared the storms, but that's beside the point. The storms keep coming. More often. It is getting warmer and wetter. Strangely enough, it's getting colder too.

> Residents along the eastern seaboard are starting to notice the storm's approach, and several million are making their final preparations. As the storm moves northward, it will meet the other two powerful winter storms, and experts say that even if "Sandy" loses momentum over land, the rare hybrid weather system that forms will wreak havoc along 1200 kilometers of the east coast of the United States.
>
> "We're talking about 50 to 60 million people who may be hit," says Louis Uccellini, a director at the National Oceanic and Atmospheric Administration to the news agency Associated Press.[9]

This was in October 2012. And early in 2014, I read about another new storm along the American eastern seaboard, which caused chaos on the highways, leaving people stranded in their cars for up to twenty hours. Cold winds were blowing, too. Both in Europe and the US.

> In American metropolises like New York, Chicago, and Boston, residents and authorities are bracing themselves for extreme winter weather. Forecasters predict thirty centimeters of snow in New York tonight, along with wind speeds up to 60 kilometers per hour, which can make temperatures fall to minus 25 degrees. Over a hundred million people – about a third of the country's popula-

> tion – are in the path of the storm, according to CNN. When the storm hit the Canadian town of Winnipeg on Tuesday, temperatures dipped to minus 31 degrees.
>
> The news bureau Reuters describes the cold as dangerous, and several schools on the east coast of the United States will be closed. On Thursday, more than 2000 flights were cancelled in the US, according to the website flightaware.com.[10]

My thoughts turn to my windproof overalls, my woolen underwear, and my boots for walking on ice. I think of my woolen mittens, my balaclava, and the big scarf I wind around my down jacket. This is gear I had to equip myself with to participate in arctic expeditions or visit good friends in Longyearbyen. My father was just as well equipped when he went out ice fishing. We have top-notch equipment and down comforters; my home is well insulated, with a cellar full of dry firewood. I like the cold. I grew up on the windswept coast at the edge of the vast Arctic. But how does the cold affect people who have never before experienced such low temperatures? People who don't have warm houses, fuel, and clothing?

"Tomorrow, people should definitely consider staying indoors if the storm continues. This is no laughing matter. The past couple of years have shown us the power of nature. We've seen how much damage it can do," says New York's Governor Andrew Cuomo to Reuters.

This is not "far away". It is here and now, and I see the images, the tragedies, and the news on my iPhone, iPad, or computer screen almost the moment they happen. This is not far away. This is reality and we know it. We know it more than well.

But does that make me a better inhabitant of this planet or a more conscientious person? When did I start thinking about the climate and sense the smell of something smoldering in my kitchen? Do I do anything? And am I doing enough?

Alarming new knowledge about ecotoxins

It may have begun to dawn on me about ten or fifteen years ago. In the early years of this millennium my focus as a journalist was on environmental pollutants in the Arctic. I visited Svalbard and northwestern Russia to document the accumulation of ecotoxins, PCB, brominated flame retardants and radioactivity in the North. It was frightening. Everywhere I went people also mentioned that a rise in temperature would make things even worse.

Senior scientist Geir Gabrielsen at the Norwegian Polar Institute took me and the television cameraman to the inner end of spectacular Kongsfjorden, on the northwest coast of Svalbard. In a rubber boat we approached the mighty snout of the glacier – a sight no words can express. Awe-inspiring beauty and glistening cold colors, glacier ice shaped by thousands of years of cold. It made us feel insignificant. We gazed with amazement and a touch of anxiety at the glacier's seaward end. We kept our distance – no cheating with the safety margins.

"This glacier is retreating so fast you can see changes from year to year. I've done research in Ny-Ålesund every summer for the past 25 years, and have seen astounding changes just over that short time in world history," said Geir Gabrielsen on that occasion.[11]

And the glacier calves as we linger in front of it, our film camera running. Suddenly huge slabs of ice break off and fall into the water with a noise like thunder. It is a spellbinding sight. Incredible and thrilling to see close up. We are there, and we see it happen. We are speechless with amazement. But it is also a warning, an alarm bell for what is happening in the Arctic. And the Arctic is a warning light for what is happening on our planet. The Arctic is the canary in the coal mine that falls silent when something is wrong with the air. What is happening in the Arctic should wake us up. The alarm is sounding. But is anybody listening?

Sometimes it appears not. Alarm bells are clanging, but still, we – the privileged people on earth – continue our disturbing race for even more consumption, even faster use of resources, even more hazardous pollution. The less privileged hope for a better standard of living, maybe on par with our own, though everyone understands that the earth cannot support it. Even now we see an imbalance that must be put right.

Poor countries do not want to pay the price for the industrial emissions that we, the wealthy, have let loose on the earth. And why should they? The industrial revolution generated unparalleled growth and prosperity. It created a rich upper class. The poor part of the world struggles with overpopulation, food shortage, health and environmental problems, and cherishes a hope for better conditions for those at the bottom of the ladder. Can we blame them for not wishing to atone for our own transgressions?

Populous countries like Brazil, India, and China are improving conditions for their inhabitants on a broad scale. We may dare to ask ourselves: at what cost? We who live a life of plenty – the richest of the planet's lords and ladies – we would rather not relinquish anything to benefit the masses. Or to benefit the earth, which is now undergoing a rapid change of regime. We see reactions at the major climate conferences; idealistic environmentalists from countries rich and poor stand outside, waving their placards. They demonstrate. We see them on TV and think that perhaps it's good that somebody is protesting. Because the conference doesn't achieve more than a fraction of the consensus and willingness to change that we had hoped it would, to safeguard everyone's future.

We know time is short.

I sit at home with my coffee cup and shake my head in resignation. The neighbors stop by and chat about their latest vacation in the tropics. I consider installing heating cables in my driveway. I really should take action somehow. But no riots or angry protesters demand change here at home. This is what writer and journalist Mark Lynas describes so alarmingly in his book *Six degrees*:

"There should have been panic in the streets, people shouting from the rooftops, statements to Parliament, and 24-hour news coverage."[12] But none of these things happened.

Arctic warning lamps begin to flash

When I read the 2005 AMAP report about ecotoxins with a sharp eye, I noticed that changes in climate were an important aspect of the problem. AMAP's responsibility is to coordinate monitoring and perform scientific assessments of pollution and climate change in the Arctic. At that time I was working on a book for young people about environmental toxins in the High North. And I began to realize that the Arctic plays a key role; that we need to learn a lot more about how nature and ecosystems tolerate the changes taking place in the Arctic, because the Arctic plays an important role in regulating climate around the globe. And the scenario became clear:

Future changes in temperature will be larger, the farther north you are on the planet. In the Arctic, temperatures are expected to rise two or three times more than in the rest of the world over the next hundred years. This is a cautious scenario. There are indications that it will happen even faster.

This was also the year when I started following the National Snow and Ice Data Center website to see what was happening to the ice in the High North. NSIDC is part of the University of Colorado and monitors both the frozen regions of the world and their role in global climate. Their website was almost as exciting as a good detective novel.

Melting ice becomes a hot topic

But it wasn't until 2006 when I started working on the World Environment Day at the Norwegian Polar Institute that I got a sense of the entire picture. The contours. Or the foreshadow of a new era. The World Environment Day posed the big question about the entire planet. As expressed in the theme and slogan for that year: "Melting Ice – A Hot Topic?"

At last somebody was looking at things from a broader perspective. World Environment Day was not just about ice melting at the poles. It was about desertification and drought, about how species disappear from the face of the earth every day. It was about the economic consequences of future catastrophes. It was about transmission of disease, about climate refugees in numbers we barely dare imagine. It was about populations fleeing, insurance payments, and wars over resources. It was about the rivers in the Himalayas, about the snow, ice, and mountains in Asia that constitute the world's water tower. It was about a new ocean opening in the north.

Yes, it was actually starting to get hot. Hotter than I anticipated. And I had been blind for so long. Or had I just been misinformed? In light of the worldwide activities in 2007, UNEP[13] director Achim Steiner said:

> World Environment Day has at its heart the empowerment of the individual citizen. The United Nations Environment Programme urges everyone to embrace this year's theme and put the question to their political leaders and democratically elected representatives: Just how much hotter does this topic need to become before governments across the globe commit to acting together?[14]

That was eight years ago, and how far have we come? Have the world's leaders understood that we must work together to solve these problems? I constantly ask myself if we are moving in the right direction, and if we are moving fast enough.

Warm and cold ocean currents meet

The following year, in 2008, I went on my first trip as a journalist in the Arctic Ocean, reporting on researchers studying ice, ocean, and currents through Fram Strait. This narrow strait lies between Svalbard and Greenland. Ice-cold water flows out of the Arctic Basin and meets the warm, salty Gulf Stream, which runs along the west coasts of Norway and Svalbard after its long journey from the Gulf of Mexico. This trip was part of the International Polar Year (IPY), and a few berths on this ambitious, six-week cruise had been reserved for journalists and photographers. The Norwegian Coast Guard vessel *KV Svalbard* had been leased for the trip. She is an ice-classed vessel, able to move through ice as much as a meter thick.

Every day I blogged on the NRK website; I took videos and photos.[15] Day and night were indistinguishable. There was always something happening on board. For long stretches of time the ship lay moored alongside an ice floe, floating with the current. That way we could measure how fast water and ice drift from the Arctic Ocean and out through the narrow strait. Protected by polar bear guards, we lugged equipment out onto the ice to take ice cores, measure ice thickness, wind and currents. The researchers' helicopter lifted and landed on the helipad above the aft deck, taking people and equipment into the icy waste. Aided by advanced equipment, the researchers measured the thickness of ice in the vast areas they flew over. It was never quiet on board; it was a twenty-four-seven project, dedicated to research day and night.

My last article described that the researchers could no longer find multi-year ice in the Arctic Ocean.[16] Multi-year ice is ice that does not melt away in the summer months, and it can be very old. It is thicker, tougher, and denser than the ice that formed during the previous winter season. Researchers estimate that the last time the Arctic was completely free of ice in summer was about 125,000 years ago. So it looks as though the old multi-year ice that has covered the Arctic Ocean for the past hundred thousand years is now melting away.

The researchers also measured what is called *albedo* – they registered how much of the light on ice-covered areas radiates back to space, and compared this with radiation from the dark waters of the open sea. White ice reflects most heat back to space; dirty ice and ice covered with ponds of melt-water reflect less. The open ocean reflects least. This means we must avoid polluting the Arctic even more, because then more heat will stay on the planet.

When the ice melts, earth's icebox stops working, so to speak. We can call this a vicious circle, because the more the ice melts, the less energy is reflected back to the universe, and the more heat remains down here with us. This is also called a negative feedback effect. And it leads to global warming.

Many read my somewhat flighty reports from the arctic ice. The blog got several thousand hits and prompted debate in the comments. When I returned from the vastness of the ice to my kitchen, I brought along a piece of ice from the Arctic Ocean to put in my freezer. Another lump of ice weighing several tons lay on display on the aft deck of *KV Svalbard*. We also brought along polar bear tracks frozen into an ice floe. The vessel was open to the public when we docked in Tromsø. People climbed aboard to meet the researchers and ask questions. It was a new era of open dialogue. Kudos to the knowledgeable researchers involved in the International Polar Year, who took the task of communicating with the public seriously. This is an important task.

Copepods and arctic tipping points

In 2009, I discovered tipping points. "Arctic Tipping Points" was the name of an international research cruise I was invited to participate in. The research vessel *Jan Mayen* sailed in the Barents Sea and along the west coast of Svalbard to the northernmost islands. We were looking to determine the range of a tiny copepod, *Calanus glacialis*, which is food for many bird and fish species in the Arctic. On this cruise, I learned not just the name but also the importance of another copepod: *Calanus finmarchicus*. This plankton is a key species in the ecosystems of the Norwegian Sea, just as its rather fatter cousin *C. glacialis* is crucial in the Arctic.

"They are building blocks in the marine food chain, and we know far too little about the ecology of the Arctic Ocean," said professor Paul Wassmann, who headed the cruise along with Carlos Duarte at the Mediterranean Institute for Advanced Studies.[17]

Copepods are a group of tiny crustaceans that live in the ocean. These little organisms had suddenly become more important than I could ever have imagined. They are zooplankton, vital sustenance for fish larvae. Copepod species are essential in nearly every marine ecosystem: they feed not only fish fry, but also krill, seabirds, and whales. But we were also searching for them because some types of copepods prefer warm seawater and others like colder seas. This means they are a natural indicator of the shifting borderline between warm and cold water.

"The warmer the water, the greater the productivity of species that can tolerate climate change. This may make Norway a winner in the climate game, since greater productivity can mean expanding fisheries," continued Wassmann.

This opinion is shared by professor Laurence D. Smith in the book *The World in 2050: Four Forces Shaping Civilization's Northern Future*. Norway may emerge as the winner in a future struggle over resources. This is not unproblematic from an ethical perspective. Be that as it may, the idea of a tipping point occupied my thoughts most of that summer.

The tipping point concept was seared into my memory by a coffee mug. It was held at an angle, to illustrate the situation. In a YouTube video, senior researcher Jay Zwally of NASA demonstrates the tipping point this way:[18] he puts a coffee mug on a table and tips it to one side. Up to a certain tilt angle the mug tips back to standing when he lets go. But if he tilts the mug too far, it falls over. The initial situation where the coffee mug tips back to standing position on the table cannot be reinstated. Zwally has illustrated a tipping point.

I write in my blog, text transmitted by satellite from the Arctic Ocean to the NRK newsroom in Tromsø:

> I believe the Arctic is now passing – or has already passed – a tipping point. And this is a belief I share with some meticulous scientists, political commentators and many climate activists. I think terms like "tipping point" and "breaking point" will soon be on everyone's lips.[19]

Is it you and I who hold the coffee mug? Do all the world's nations hold it? When will we reach the tipping point? Are we there now? Tomorrow? In ten years, or fifty? Can we find a way to stop the process and reestablish balance? The questions buzz in my head. What if the world is already entering a new phase? Can we reverse the process we have helped set in motion? An absolute majority of the world's climatologists have no doubt: human activities contribute to climate change.

It is beginning to dawn on me that things are really serious.

"The Arctic is often called the canary in the coal mine for climate change. As a sign that the climate is warming, the canary is now dying. It's time to get out of the coal mine," said Jay Zwally.[20]

Highest level of greenhouse gas in 800,000 years

Back home, the media debate still focused on whether climate change was caused by humans or not. The discussion raged furiously. Fine. That meant we didn't have to take in the chilling fact that the clock was ticking faster than most researchers had dared to believe. That the models were wrong. That many calculations underestimated how soon the Arctic Ocean would be ice-free in summer. Researchers used to claim that we would have an ice-free Arctic in summer in 2100. Now the date is being moved closer and closer. Ice-free by 2050, say some – and the most daring say 2030. Or earlier. The scientists had not been exaggerating when they warned us; on the contrary, their models had been unable to predict how quick the process would be.

The United Nations' Intergovernmental Panel on Climate Change has established that the climate change we are now experiencing is with 95% certainty caused by humans.[21] In 2010 the certainty had been a bit lower. Still, this sends an important message: if we contribute to what is happening, we are also obliged to stop when we see the consequences. Ninety-five percent is certainty enough. It is more than enough to convince me.

I have no doubts. If anyone told me that I had a 95% likelihood of getting cancer if I did not change my lifestyle, I would change it. Now we know what is happening to our planet – Tellus, Gaia, Mother Earth. So I have to wonder why we hesitate. I wonder if despondency has rooted us to the spot. Or is it just denial? The facts are unmistakable.

"The atmospheric concentrations of carbon dioxide, methane, and nitrous oxide have increased to levels unprecedented in at least the last 800,000 years."[22]

And another fact is that I began to worry. I thought about it when trying to sleep; I saw several movies about it. I met Al Gore, who showed up in Tromsø to present his opinions on climate. It was alarming and interesting, regardless of how you might feel about the former vice president and climate policy in the United States. He could hold a listener's attention. Once again I watched his movie, *An Inconvenient Truth*, and must admit that it did indeed contain more than one inconvenient truth. I read my old newspaper clippings and felt a certain anxiety about what we are doing to the planet and the thin layer of air that surrounds us.

On May 13, 2008, the Norwegian daily paper *Aftenposten* writes: "The concentration of CO_2 in the atmosphere is now record high, and scientists are beginning to fear that the earth will lose its ability to absorb CO_2, according to *The Guardian*. The paper refers to recent measure-

ments from the Mauna Loa Observatory in Hawaii, published on the website of the United States' National Oceanic and Atmospheric Administration."

A chilling dip in icy waters

I am invited to participate in another research cruise. I pack my woolen clothes, footgear, overalls, protective face cream, and all the cords required to charge my camera, cell phone clock, iPad and computer during the trip. An electric toothbrush, too. Strange how many devices and electric cords modern people need. I'm traveling with the new ICE Centre of the Norwegian Polar Institute on their first ICE expedition with the old converted fishing boat, now the research vessel *Lance*, which is ice-classed and easily transports us to nearly 82 degrees north. Once again work on board runs twenty-four-seven; it is light all the time at these latitudes, and a polar bear comes right up to the hull on a spectacular visit.

Melting ice in the Arctic has become a hot topic (Photo: Sylvi Inez Liljegren)

Systematic measurements are taken from Rijpfjorden in the north of Nordaustlandet right up to 81.5 degrees. The ice is not very dense and we sail north essentially unimpeded.

Before we head for home after three weeks in the ice, researchers from around the world take a dip in a lead in the ice, a rope around their waists. One by one, they plunge into the same icy ocean where scientists and teams of divers have been collecting samples for three weeks. They are making sure they will never forget this cruise. When the water is –1.8°C you feel the cold, even if you are bundled into warm clothing at the edge of the lead with your camera at the ready.

And then we head home along the magnificent glacier Blåsvellbreen on the south coast of Nordaustlandet; the researchers have fresh data about ice conditions, temperature changes, biological processes and oceanographic parameters, and I have new insight, photos, and films. We deliver a story for the television news program *Dagsrevyen* with photos from research over and under the ice. Marine biologist Haakon Hop from the Norwegian Polar Institute expresses concern for the teeming life forms under the ice and the threats the future holds for these organisms if the ice in the Arctic Ocean disappears in summer.[23]

I didn't realize the community under the ice was so rich and diverse. For the first time, I consider buying an electric car. I decide I like being in the Arctic. Maybe I would have been more useful as a scientist than as a journalist. And I mumble to myself the most important message:

"There should have been panic in the streets, people shouting from the rooftops, statements to Parliament, and 24-hour news coverage."[24]

The tragedy of the commons

I see we have become part of what the American ecologist Garrett James Hardin called "the tragedy of the commons". Hardin was particularly interested in overpopulation. He is described as "an ecologist who called attention to the damage that innocent actions by individuals can inflict on the environment".[25]

The tragedy of the commons arises when one or more of those who use a common, harvest more than their share of the growth of a renewable resource. The classic example is overfishing, where one or more fishermen land more than their share. The same might be said about reindeer grazing on Finnmarksvidda, where one or more herders allow more reindeer to graze than the communal pastures can sustain. Overfishing and overgrazing eventually destroy the resources, and everyone ends up worse off – including those who helped themselves with both hands.

I reread the letter Chief Seathl of the Duwanish people in the state of Washington wrote to President Franklin Pierce in 1855. The American authorities wanted to take over tribal lands. The chief wrote:

> How can you buy or sell the sky, the warmth of the land? The idea is strange to us. If we do not own the freshness of the air and the sparkle of the water, how can you buy them?
>
> We only decide about our own time. Every part of this earth is sacred to my people. Every shining pine needle, every sandy shore, every mist in the dark woods, every clearing, and every humming insect is holy in the memory and experience of my people.
>
> We know that the white man does not understand our ways. One portion of land is the same to him as the next, for he is a stranger who comes in the night and takes from the land whatever he needs. The earth is not his brother, but his enemy, and when he has conquered it, he moves on. He kidnaps the earth from his children and he does not care. He leaves his fathers' graves, and his children's birthright is forgotten. His appetite will devour the earth and leave behind only a desert.[26]

The new north

Ecological thinking is nothing new. But the oil and gas companies are now looking hungrily toward the promised land in the north. Because when the ice melts, they envision new trade routes and new possibilities to extract raw materials. But the earth does not belong only to the rich, the powerful, the large corporations – those who make money without sharing; not to property magnates and companies that don't want to clean the smoke they send up their chimneys, that don't want to be responsible for the waste they discard in secret, that don't want to remove the armaments and radioactive trash they have already scattered around the Arctic.

New areas are now opening up for oil and gas production, areas that have previously been covered by arctic sea ice, areas that can be exploited to generate new wealth and expansion, but simultaneously release carbon compounds that will not benefit the atmosphere or the environment. The ice melts, revealing "The New North", which some equate with "The Promised Land". Climate conferences ostensibly intended to draw up new strategies for the world look more like parodies, reaching little or no real consensus. The Kyoto protocol has become outdated. In China, coal-fired power plants pop up like mushrooms. On bad days, Beijing is wreathed in worse smog than Los Angeles, and many Republicans in the United States distrust any sanction that might impede commerce. Inuit children in Greenland already have high concentrations of lead and mercury in their blood, people living along the coasts can no longer eat cod liver or seabird eggs, and women who nurse their babies must live with anxiety at the thought of the environmental toxins that leave their bodies along with their breast milk.

It looks as though a conservative United States will not support Obama in a fight to avert an approaching climate crisis. This is one of many paradoxes. But Obama, somewhat unexpectedly, has declared that next year's budget will include a "Climate Defense Fund". He himself has stopped smoking for the sake of his health. If he also manages to cut CO_2 emissions for the sake of the world's health, we will have moved forward. In Russia, environment activists are imprisoned and are considered a threat to the nation. Images from major cities like Beijing and Tokyo routinely show people in face masks; what those citizens are thinking behind their masks nobody can tell. But the situation doesn't look bright.

Warmer oceans give cold spells

In the midst of everything, the cold spell confused me. Wasn't it supposed to get warmer when temperatures on earth increased?

"Last winter, parts of Europe were stunned by excessively cold weather. In countries like Poland, several people froze to death. Nobody was prepared for record low temperatures and heavy snowfall over such a long period of time. Researchers believe they now understand what causes these sudden cold spells in Europe. The reason lies far in the north, in the icy regions around the North Pole. A strong wind has 'always' kept the cold air still over the Arctic Ocean. Last winter, the jet stream broke up and ice-cold air masses spread down over Europe," says Jan Gunnar Winther at the Norwegian Polar Institute to the daily paper *Aftenposten*.[27]

It is going to get wild and wet. And maybe both warmer *and* colder. Because someone has pulled the rug out from under us and upset the entire system. Can it be restored?

Winther also mentions meteorological changes as one of several factors we know too little about: "We know that the future climate will be wilder and wetter. We also know that it will be less predictable. But many things are happening, and we need to gather more knowledge about them. Much of what is going on in the climate in the north will have consequences farther south." He doesn't exclude the possibility of new surprises that will bring lots of snow and cold in the future.

Some sketch a worst case scenario

Since greenhouse gas emissions continue to increase, we are warned that global temperatures may rise as much as six degrees Celsius over the next century. This could cause radical change on earth. In the documentary *Six Degrees Could Change the World* the British writer Mark Lynas and climate experts present their opinions about what types of effects a temperature increase of that size would have on the earth.

Even if release of greenhouse gases were to stop immediately, the concentrations already present in the atmosphere will cause global temperatures to rise by one-half to one degree. But what if the temperature should increase by yet another degree?

According to Mark Lynas, author of several books on this theme, the changes will no longer come gradually. Here is an excerpt from a write-up about the film *Six Degrees Could Change the World*:

> The glaciers in Greenland will begin to disappear, and with them some low-lying islands. At three degrees higher temperature, the Arctic will be ice-free all summer, the Amazon rain forest will begin to dry out and extreme weather patterns will become the norm. At four degrees higher temperature, sea levels will rise dramatically. Subsequently there would be a transition involving climate changes, if global temperatures rise yet another degree. Parts of what were once temperate zones would become uninhabitable and people would fight over the world's remaining resources. The sixth degree can be called a doomsday scenario, when oceans become marine wastelands, deserts expand, and natural disasters become more frequent. If we do nothing to avert this threat, when will we reach the point where we are longer able to stop global warming?[28]

A shopping-free year and cognitive dissonance

I put a one-year moratorium on purchase of clothing, footgear, furniture and other non-essentials. Basically I find it easier than expected. But I continue to buy good food and wine for weekends. And I continue to drive to work. I must admit I envy the colleague beside me at the intersection in his new electric car, and how it glides away silently when the light turns green. But I can't afford a new car, and my 20-year-old Rabbit is still going strong, though it emits exhaust I'd rather not think about. At least not yet.

They say it helps. That we as individuals can actually make a difference by the way we manage our house, our food, and our car. But I suffer from cognitive dissonance. By that I mean that I know what I *should* do, but I don't do it in real life. And that conjures up in me an inner storm I cannot put to rest.

I – along with many of the world's scientists, writers and journalists – must also consider how I communicate my knowledge. And my worry. Maybe we need to find ways to converse about what frightens us – new ways, so that the conclusions we arrive at actually contribute toward positive change.

An old classmate of mine who studied biology now travels the world as a senior researcher, giving lectures. And I write blogs from research cruises. When I come home after a three-week cruise, most of what I have written has already been read, and new issues have captured people's attention.

Am I of any use? Does it help to describe climate change when the world appears to be going full speed ahead on the wrong track? I sit at home in my kitchen with a cup of coffee after my latest cruise in the Arctic. And I don't know what to say and do.

Why don't I use every single day to stand up and tell people what I've seen? Why can't I live in accordance with the reality that is becoming so clear to me? What is it about us humans – that we are unable to grasp the seriousness of this threat that hangs over each and every one of us?

In 2012 the world hit a new record, according to the website of the National Snow and Ice Data Center: "Arctic sea ice appears to have broken the 2007 record daily extent and is now the lowest in the satellite era. With two to three more weeks left in the melt season, sea ice continues to track below 2007 daily extents."[29]

We beat all previous records for melting of arctic sea ice. This is a negative record. And probably not the last one we will see in years to come.

A whiff of smoke

An e-mail arrives from the *Climate Reality Project* in the United States. Today they write: "We are powering social revolution on climate change". A social revolution for climate change? At least it's a start.

> Most people don't know that emissions from coal-fired power plants are completely unregulated. These emissions aren't monitored and it costs us. Five years ago, the American Congress had a historic opportunity to use market forces to limit carbon emissions, but after the bill passed the House, it was voted down in the Senate.
>
> Since then we have seen the disturbing consequences that the climate crisis has to offer – from a drought that covered 60% of our nation to Superstorm Sandy, which wreaked havoc and cost the taxpayers billions, from wildfires spreading across large areas of the American West to severe flooding in cities all across our country – we have seen what happens when we fail to act.[30]

We have a long road ahead of us. What is it we are beginning to discern? A wish to retain our freedom to go full speed ahead without restrictions – even if the road leads to hell? I feel complicit, yet I know a bad conscience won't help. But nobody asks anymore who owns the freshness of the air and the sparkle of the water. And I have become a cog in a machine that is moving in the wrong direction, and I no longer know if my little voice can be heard.

But if you sit working or reading or chatting with your friends and the kitchen catches fire, how do you act? Do you stay put and continue doing whatever you were doing? Do you look the other way and pretend it isn't happening? Do you run out of the house screaming and expect someone else to extinguish the fire for you?

We know what's happening: the kitchen is smoldering, something has caught fire. But we're reluctant to take it all in. We continue as before, despite the whiff of smoke and even though the wallpaper is curling at the edges. I sit there with my coffee cup and send a text with the word "NEED" and feel that I've done my part once again.

Soon enough, the wiring over the stove short-circuits. And poof! There goes the electricity.

I can't charge my cell phone anymore. No more news programs. No more well-intentioned texts to donate money for those in need. Was it in the Caribbean or Florida, or maybe in the Philippines? These things happen more and more often. Maybe it was all just part of a movie I saw recently? Or propaganda from some hysterical environmental group? I can't quite remember which messages arrived most recently; there were so many right before the electricity blew and my PC shut down. Was Europe freezing to death in a new cold snap? Had the Netherlands been flooded by the Rhine? Has another platform succumbed to ice in the Barents Sea? What's that you're saying? I should I have thought of this earlier?

At any rate, I have to get out of the house. I won't have time to finish my coffee. Something smells badly scorched and the smoke makes my eyes run. I'm worried about potential damage to my newly polished fingernails, my best silver, and my new skiing gear. It's too late to save the kitchen, at any rate. I realize that now.

POSTSCRIPT

A fire brigade. What we need is a fire brigade. To put out the fire before the flames rise too high. There are thousands of them out there, maybe millions. Volunteer organizations, environmental activists, secondhand stores, anti-consumption groups, church leaders, men and women who ride bicycles to work instead of driving, people who recycle trash, grandmothers who teach their grandchildren to set up a loom, cut strips of worn-out cloth and weave rag rugs, people who turn off the water while they brush their teeth because they realize that we live on a planet with limited resources.

A disposable planet?

The planet needs people working in research and development to advance technology and renewable energy, who write books describing 365 things you can do to live greener every day of the year. People who take pains to choose fair trade products, who wear hand-me-downs, and turn off the light for the sake of the planet – not just one hour a year, but every day. We need the girls who jump off the fashion bandwagon and choose sensible, sustainable clothing, who make it trendy to wear used clothing. And we need you, over there – yes, you – who sort your trash and recycle what you can.

If I am anything like the average Norwegian consumer, I throw away 51 kilos of food per year. That is hair-raising. But there is almost no end to what you and I can do to become part of the world's fire brigade. Not just the fire brigade that puts out fires, but also the proactive part of the brigade that does inspections and prevents new fires.

At the same time, major changes are needed. It is people with power and influential politicians who must act nationally and internationally. And they need daily reminders about how important it is to act. But I also believe in trend-setters, celebrities, and musicians. It isn't the first time in history that big stars have said important things from the stage to a global audience eager to listen. Mandela at his 46664 concerts and Sting with his concern for the rainforest are good examples. That makes an impression and makes us all listen together. And listen to each other.

"We are all Africans," said Mandela in Tromsø in 2005; we are all citizens of the earth and must aspire together.

Changes in climate are not somewhere in the future. They are already here, making storms on the coast of India less predictable, expanding deserts so women in Africa must walk even farther every day to find water for themselves and their families. If the Indian monsoon changes, it will affect nearly a sixth of the world's population. If sea level rises, climate refugees will become a worldwide problem, one we cannot simply turn our backs on.

Some will say we're on a first class journey to hell. Whatever. But others have lit torches and raised their voices, and soon new beacons will burn on mountaintops all around the world. For peace, for coexistence, for an environment and a climate where our children, grandchildren, and great-grandchildren can continue to live. More people join the fire brigade every day. Every bit counts, and your acts contribute. Everything is better than apathy. And in the words of Confucius, it is better to light a candle than to curse the darkness.

Longyearbyen in the magical blue light of late January (Photo: Diana Snibsøer)

A look into the future

We posed some questions to five selected citizens of Longyearbyen

What are your thoughts about the future of Svalbard? Are you concerned about climate change?

Odd Olsen Ingerø, Governor of Svalbard

I've noticed that the message from IPCC has become increasingly clear. They expect a more rapid temperature increase in the Arctic than elsewhere in the planet. The trend we are now seeing is a warmer climate and less ice in the fjords and around Svalbard. If that trend continues, it is worrisome in view of the changes it will cause in the natural environment.

We already see signs that new plant and animal species are gaining a foothold in Svalbard, both on land and in the ocean. That isn't troublesome in itself, but it can be detrimental if it perturbs the balance in our ecosystems.

Changes in ice conditions mean that we already see increased shipping across the Arctic Ocean. The risk of accidents and harm to both people and the environment will increase as a consequence.

Viljar Hansen, student

I strongly believe that Svalbard will become an increasingly important part of the world in coming years. We see how our part of the Arctic is attracting attention, not just from our closest neighbors, but also from the United States, China and countries on most of the world's continents.

We also see that the Northeast Passage, which is ice-free more often nowadays, will be used more frequently. Svalbard can become an important center for rescue and oil containment operations. I think that will give Svalbard an important position in the future.

Norwegian presence in Svalbard will be a crucial factor if Norway is to benefit fully from activity in the Arctic. But it's a double-edged sword. The reason why focus is shifting to the Arctic and Svalbard is that oil wells in other parts of the world are running dry and oil and gas extraction in the Arctic is looking more and more feasible.

I'm afraid of the potential consequences of oil operations near Svalbard, so I think research aimed at making oil operations as safe and environmentally responsible as possible is important, in addition to research on renewable energy sources, which I hope are part of Svalbard's future.

Research shows that global climate change will be visible first in arctic and antarctic regions. I've noticed that conditions for snowmobiling and the weather in Svalbard are not as stable as they used to be. If we can already see that the snow is melting earlier than usual in Svalbard, that's cause for concern. In other words, I'm really worried about climate change.

I hope people will be able to enjoy the fantastic natural environment and resources in Svalbard for many generations to come. For that reason, I hope all future activity in Svalbard – from tourism to commerce and maybe extraction and use of fossil fuel and renewable energy – will be carried out with the least possible negative impact.

Anita Johansen, Union representative, Store Norske Spitsbergen Kulkompani

I think the non-Norwegian segment of the population in Longyearbyen will grow in the future. This will pose great challenges in terms of the rights children are expected to have in Norwegian society. Any major shift in the relative proportions of Norwegians and foreigners can raise questions about sovereignty, particularly in the case of "aggressive" Russians and competition for resources such as fish, minerals, oil, and gas.

I expect to see increased Norwegian activities in research and education, reduced mining, and stagnation in and restrictions on tourism. Initiatives to expand research and education are positive, but mining operations still play an important role in keeping families with children here, and ensuring that the day-care centers and school in Longyearbyen survive.

Mining is also the main guarantor of continuity in a town that has an average population turnover of about twenty percent.

I envision increased focus on environmental issues. This will probably lead to more conflicts between commercial interests and central authorities. I think this can also prompt other parties to the Svalbard Treaty to challenge Norwegian environmental legislation. The Treaty brings countries together in a completely different way than any other international negotiation forum, and I think the struggle for resources will increase dissension between them.

I think climate change both poses major challenges and offers new opportunities in and around Svalbard. In my opinion, Svalbard has become both a *national* and to some degree an *international* symbol of climate change. I also think the negative predictions about climate change get a lot of the attention. Obviously nobody wants the polar bear and other animals to disappear, but I'm not sure how realistic that actually is. So I guess I'm not excessively worried about climate change.

Alexandra Anna Smyrak-Sikora, PhD student, UNIS

Climate change is a topic everyone cares about. To discuss any ethical dilemma linked to climate, we must first understand that climate is not a static phenomenon. Climate changes, and we cannot expect to keep the climate unchanged or frozen in its current form. Climate has changed considerably throughout earth's history.

Even if we only look back over the past millennium – which is but a moment in the billions of years of the earth's history – we see that the climate was warmer than it is now between 1000 and 1200 AD. The climate was and is changing: it is variable by nature.

Recently we have experienced significant population growth. The earth's population has now passed seven billion. That is twice as many as forty years ago, and it affects our environment. Humans pollute their surroundings by discarding more trash and using more energy. A new cell phone has become more important than the air we breathe.

Svalbard is the place where we can easily observe interactions between climate and cryosphere. The inhabitants of Longyearbyen have negligible impact on global climate processes. But we affect the quality of the air we breathe in our own town, and we produce the waste that surrounds us. If we wish to be a community with little or no impact on the environment, we must reduce our emissions – of exhaust, for example.

How many of us will give up driving in Longyearbyen? Can the Arctic continue to be an attractive destination without the snowmobile excursions that allow us to enjoy nature and look at polar bears on the east coast? How much can we cut air travel? Can we limit tourism? These are a few of the questions that have no obvious answers.

Terje Aunevik, General Manager, Pole Position Logistics

I strongly believe in a good future for Svalbard. At a time when the world is looking north, Norway and Longyearbyen have a unique opportunity to develop an arctic capital with forward-looking research institutions and commercial enterprises that work together to create values and solve tomorrow's problems – be they in energy, communication technology, tourism, shipping, mining, industrial logistics or some entirely new fields of endeavor.

I am always on the lookout for symbiosis; creating links between different environments can give astounding results. To achieve that, we must dare to think outside the box, freely and frankly, and constantly search for possibilities instead of focusing on limitations. So far, we haven't been good enough at that in Svalbard, but things are changing. This is not just a necessity, but also incredibly inspiring and fun!

There are many indications that climate changes will be more visible in the Arctic and in Svalbard than in most other places. So far, we know little about what the ultimate consequences will be, apart from less sea ice, milder weather and shrinking glaciers. Sometimes I feel that the fact that these changes are so visible in the Arctic leads to a dead end in the climate debate, that enacting extra strict management regulations up here gets us side-tracked. Because if we are to believe the theory that humanity's CO_2 emissions are the most important driver of climate change, the solution requires a lot more than local protective measures in Svalbard.

What possibilities do you see for renewable energy versus fossil energy?

Odd Olsen Ingerø, Governor of Svalbard

Energy production is way outside my competence. That said, it's clear that the relationship between emission of CO_2 from fossil fuels and climate change requires us to think innovatively. Making our energy production CO_2-neutral would be a natural continuation of the exacting environmental standards that have been set for Svalbard.

About ten years ago, the energy company Statkraft investigated whether it would be possible to use the wind to generate power in Svalbard. Those studies showed that there was not enough wind in the places where measurements were made. The situation may change, and technological advances are rapid in this field. I note that the mining company Store Norske is experimenting with solar power at the mine in Svea. It will be exciting to hear what conclusions they draw. I believe renewable forms of energy will force their way into our future, also here in Svalbard.

Viljar Hansen, student

I hope and believe that Svalbard's future is renewable. Coal mining has been and is still tremendously important for Norwegian presence in Svalbard, but extracting fossil fuels is not sustainable in the long run. I think that with research we can find other – renewable – energy sources in Svalbard.

Svalbard has lots of weather: we who have lived here know that! I am confident that energy from waves, wind, and other renewable sources can work quite well in Svalbard. But a political will to invest heavily in research on alternative energy is a prerequisite.

Norwegian presence in Svalbard depends on establishing new major employers that are not based on coal, and that research and tourism continue to grow. This is required to sustain the basis for Norwegian presence and to maintain the Svalbard Treaty.

Anita Johansen, Union representative, Store Norske Spitsbergen Kulkompani

Parts of the western world are focusing on this, but what we have noticed is that activity levels depend on the economic situation. Prominent analysts predict that fossil energy sources will continue to dominate for many years. I don't think there will be any serious investment in renewable energy sources until fossil energy becomes scarce. Once again, economic factors play a decisive role.

I think the trends in Svalbard are the same as elsewhere. And here we can't strike an agreement with an adjacent county to help each other out if there's not enough wind power. That means we will have to have access to both kinds of energy for a long time to come.

There have been attempts to measure winds, but apart from that, there has been little discussion about using renewable energy in Svalbard. We have ready access to coal, and our infrastructure is based on energy from coal.

More should be invested in carbon capture and storage and similar techniques, which would be fantastic to export to a world that relies so heavily on energy from coal but does little or nothing to clean up emissions.

Alexandra Anna Smyrak-Sikora, PhD student, UNIS

The demand for energy is growing worldwide, particularly in developing countries. Research publications show that renewable energy is inadequate to cover global energy demand. Industrial and economic advances go hand in hand with increased energy consumption.

I wish we could live in a world based on renewable resources, but I don't see how that is possible. The renewable resources can't keep up with the demand for energy. It isn't fair to blame the oil industry. We have to look at our own everyday actions. Over the last twenty years we have cut the amount of energy it takes to run a refrigerator, but how much more do we need for our cell phones? How much do we fly now compared to ten years ago?

Terje Aunevik, General Manager, Pole Position Logistics

Alternative and renewable energy sources are the future. But it's crucial for us to understand the magnitude of the world's energy needs, both today and – not least – in years to come.

For example, there are estimates that the middle class in Asia will grow from 500 million to 3.3 billion people in the next thirty years, which will drive up energy demand even more. This has to be seen in relation to the energy density in fossil fuels, and in that perspective it is naïve to believe that renewable energy can replace fossil energy any time soon. In this context, Norwegian gas will be a part of the solution, though that doesn't mean we should claim it is our contribution alone.

It will be more important than ever that we invest in carbon capture and storage during the transition to renewable energy. Here in Longyearbyen we have a unique opportunity to serve as an arena for research and development. We are the only place in Norway with geological prerequisites that allow us to test this type of technology on land. And our research and education institutions offer the expertise required to develop and refine it.

Longyearbyen should be a showcase. If we succeed in this, we can talk about a moon landing with our credibility intact.

How can commercial enterprises develop in harmony with the Arctic?

Odd Olsen Ingerø, Governor of Svalbard

This is a broad and difficult question and others may have much better answers. As I have understood it, what happens in Svalbard is not the main threat to the Arctic. Svalbard is affected by pollutants transported from far away and by greenhouse gas emissions in other parts of the world. Still, we should obviously do what we can locally to limit CO_2 emissions and other pollution.

Where Svalbard is concerned, the important thing, from my point of view, is that commercial enterprises develop within the established framework. Norwegian Svalbard policy and legislation are a good foundation for protecting the unique assets we have up here in the north. Paragraph 10 of the Act for Protection of the Environment in Svalbard establishes the principle that "activities in Svalbard must use whatever technique that puts the least pressure on the environment". The Svalbard Environment Act is one of the world's strictest in this context. Perhaps others should copy it?

Viljar Hansen, student

I think it's important for businesses to develop so Norwegian presence in Svalbard can continue and be productive. I think tourism, research, and energy will continue to be the main sources of income in Svalbard in the future.

The focus on environmentally friendly tourism will be crucial if we want to make the best possible use of our islands. Tourism must not only grow; it must also be made more environmentally friendly. And I believe that research will be the most important economic activity in Svalbard in the future. As I see it, developing and investing in UNIS plays a decisive role for Svalbard, since it will not only strengthen our presence in Svalbard and support local commerce, but also determine how well we manage Svalbard in the future.

I think commercial interests in Svalbard can develop in a way that does not harm the Arctic, as long as investors – truly ambitious ones – focus on the right things. And it's research and tourism I have in mind.

Where coal mining is concerned, I think it will have to be phased out as quickly as that is feasible, and then we must have other energy

Will the University Centre in Svalbard (UNIS) and other research institutes be the most important pillars of Svalbard's economy in the future? (Photo: Eva Therese Jenssen)

resources ready. *Until* this is feasible, coal mining must be phased out as safely, and with as little environmental impact as possible. Once again, research will be crucial.

Anita Johansen, Union representative, Store Norske Spitsbergen Kulkompani

That depends on what you mean by "in harmony with the Arctic". If the Arctic is supposed to be untouched, there is little scope for any kind of commercial enterprise. All human activities and all commercial ventures leave traces. But that would contradict the Svalbard Treaty, which regulates the right of co-signing nations to engage in economic activities.

I think it will be a balancing act. We need legislation that holds commercial enterprises responsible, so they must minimize any detrimental effects their activities may have, and legislation that does not provoke co-signing nations to challenge Norwegian sovereignty.

These prospects may sound unlikely, but when fish stocks move farther north, and if the ice disappears and makes more resources available for exploitation, then I think the fight for resources will intensify.

Alexandra Anna Smyrak-Sikora, PhD student, UNIS

To preserve the Arctic in its present condition, basically we should all just pack up and move. But seriously, our presence will always affect the environment and put pressure on nature.

We have to realize that everything we do leaves marks on our surroundings, and we have to take responsibility for them. We have a direct influence on lots of other factors besides the climate: exhaust emission, waste production, and – something that's rarely mentioned – noise pollution. As a community, we can decide which activities we want to allow here in Svalbard.

We can discuss whether we should support the tourist trade that focuses on observing pristine nature rather than "shopping tourism" that focuses on Svalbard as a tax-free zone. The quality of our local environment is intimately related to us and the choices we make.

Terje Aunevik, General Manager, Pole Position Logistics

In this context it's crucial for us to think globally, from a broad perspective. It isn't just a paradox, but also a source of frustration that the Arctic has become a stage for the western world's guilty conscience about nature and the environment.

It's amazing what protective measures some have managed to drive through based on a misconception that those measures will have some kind of effect on global environmental problems. To put it another way: a measure against wear and tear caused by people walking on the tundra leads us astray when we must solve the major challenges of the future. Measures like that put a damper on people's enthusiasm for discussing important issues concerning the future.

These matters interest me strongly. When I was younger I was deeply involved in the environment; in fact the government took me to court when I refused to do military service based on arguments related to use of resources and the environment. I am still deeply committed to the environment, but I'm disturbed when it gets elevated to an almost religious status. Either you are one of the believers – and in that religion you are expected to consider all human activity as a threat – or you are a doubter. And if you are a doubter, you will go to hell.

My view is that humans are a part of nature; that means that we leave marks, we affect our environment, we harvest what it provides, and we need nature to live our lives.

The trend we now see is to try to disconnect humans from nature, and that nature should be untouched. We will achieve nothing if we think we can protect the Arctic by putting it under a cheese dome. We have to think globally.

If we believe that CO_2 is the main problem, then as far as I can see there's only one way to limit CO_2 emissions, and that is to make it expensive to emit CO_2. Then other energy sources will emerge.

The discussion about letting Norway's oil and gas reserves remain underground is meaningless as long as there is strong global demand that will be met anyway, and perhaps using production technology that costs even more energy.

The commercial enterprises that are currently developing in Svalbard *are* in harmony with the Arctic. They operate responsibly, within an established set of strict rules that do not need additional tightening. It is more important to work on the emission sources that actually contribute to the environmental challenges that threaten the Arctic – and they are not local. This is why a broad perspective is incredibly important.

What ethical dilemmas do you see in developments in the Arctic?

Odd Olsen Ingerø, Governor of Svalbard

The greatest ethical dilemma we all have to deal with is the monumental use of resources and energy by people in our part of the world.

The fact that most of the world's population has a lower standard of living than we do is a prerequisite for our lifestyle, because the world doesn't have enough resources to raise everyone to our level. If everyone on earth were to live and consume the way we do here in Norway, it wouldn't be sustainable.

The ethical dilemma that is most prominent where developments in the Arctic are concerned, is our obligation toward future generations and the consequences our lifestyle has on their future. By that I mean both their opportunity to access resources, and their opportunity to experience the natural environment we ourselves prize so highly.

When you get right down to it, there's only one solution to these challenges. Politicians, especially in Norway and the western world, must take a stand, be resolute, and make decisions that will limit emissions – even though that will have consequences for individuals.

Viljar Hansen, student

I see a huge ethical dilemma in the fact that Norway on the one hand wants to join the race to exploit oil and gas in Svalbard, since that would boost the economy, but on the other hand must ensure that development is sustainable and environmentally responsible. This pits environmental values against economic gain. Sadly, history shows that economic gain usually comes out on top. The same dilemma faces other countries that compete for the same resources.

Another dilemma is whether we should focus all our research efforts on renewable resources, be the first to find profitable and sustainable solutions, and thus benefit from our knowledge and competence. Here we should not hesitate, but go straight to work on this ambitious project – and for once not allow short-sighted economic gain to stand in the way. Because the world needs this right now. Svalbard needs it now.

Anita Johansen, Union representative, Store Norske Spitsbergen Kulkompani

Balancing between economic and environmental interests is what I see as the greatest ethical dilemma. Who is going to regulate this, and how should it be regulated? Who should have access to resources, and on what conditions?

I think the greatest challenge will be finding solutions that are acceptable to all the parties involved in the Arctic. So far we haven't seen any major results come out of attempts to negotiate international agreements on environmental regulations.

Alexandra Anna Smyrak-Sikora, PhD student, UNIS

As a geologist, I don't view the Arctic as a closed system that should be isolated. I see the Arctic as an integral part of our planet. The advantage of the Arctic is the extreme conditions that have kept human influence to a minimum.

It's difficult to separate ethical dilemmas in the Arctic from global issues and conditions for the entire planet: rain forests, the earth's lungs, are just as important as the ice cover.

When we discuss climate, we have to look at the entire global context, all the processes that make human life possible.

We are changing our planet, but nature will adapt to new conditions, just as it has for billions of years. The question is whether we humans can adapt in the same way.

On a local level, I think we could have more discussions about how we want our community to develop, and in what direction. We can't close off the Arctic, but we have a responsibility to establish rules for those who want access. The development of Svalbard and the tracks we leave behind are our responsibility – both the local community and the authorities.

Terje Aunevik, General Manager, Pole Position Logistics

There are several dilemmas, and in my opinion some of them are counter to what are traditionally considered ethical dilemmas in the Arctic. For instance, there is obviously a lot of emphasis on the reduction of sea ice in the Arctic Basin. In Norway this is usually only discussed in moral terms, and exclusively from a doomsday perspective, whereas I think we should also view the scenarios from the perspective of opportunity. If we do that, we will find effects that are clearly beneficial, not least for the environment.

If the current trend continues, it will eventually become possible to sail across the Arctic Ocean – that will save lots of fuel and cut emissions. But many are nonetheless troubled by the mere thought of maritime activity in the Arctic. The visual image people want to have of the Arctic overshadows reality and the positive effects that may lie before us.

We see tendencies to the same thing in Longyearbyen. Our power plant is overdue for replacement; we have an opportunity to think ahead and re-create Longyearbyen as a carbon-neutral community. Instead, we keep patching up the old plant to satisfy emission requirements on the short term.

Our greatest ethical dilemma concerning development – be it in the Arctic or the tropics – is our hopelessly immoral consumption. It doesn't take rocket science to understand that it is impossible to bring the rest of the world's population up to our materialistic standard without a negative impact on the environment. Our extreme consumption, in all areas, is not sustainable. It frightens me that we seem incapable of focusing the debate on this aspect.

In many ways it seems the focus is on idling engines, when it would in all likelihood have a greater effect if we all ate less meat.

Another major ethical dilemma for Norway is how we invest our revenue from exploitation of fossil fuel. For some reason we seem to be awfully proud of our Oil Fund (officially called the Government Pension Fund Global). In my opinion we should discontinue the Oil Fund in its present form and my reasons are as follows:

The future is in renewable energy. We must immediately invest all possible resources to keep our country's position as a leading energy provider. We won't do that by simply pumping up oil and gas and investing the money in controversial stock and business districts in London.

We must position ourselves as soon as possible for the future energy market. That means we have to be best at all kinds of power: hydroelectric, solar, wind, wave, and other new energy sources. We have to be best at storing and distributing energy. And we have to build a society that uses the best of these technologies. If we achieve that, we are creating something of value for future generations.

I find it worrisome that we save our assets in a fund with the chief objective of financing future pension schemes instead of investing to stimulate Norwegian value creation. That is pure egoism. The values in the fund are and remain paper values. In view of the turbulence on the global financial arena in recent years, we should have learned that such investments are highly uncertain. This is not sustainable, and does not create real value for future generations.

Because that's precisely what it all boils down to, as I see it. What kind of society do we want to pass on to our descendants?

The abandoned Russian settlement Pyramiden, at the inner end of Isfjorden (Photo: Leif Magne Helgesen)

Ex machina

by Kjartan Fløgstad, writer

At the inner end of Isfjorden in Svalbard lies Pyramiden, which in this case designates both a mountain and the deserted Soviet model city that lies at the mountain's foot. It sounds incredible – it *is* incredible – but all through the Cold War, Soviet Russians and Norwegians lived peaceably side by side on this group of islands near the North Pole. The collapse of the Soviet planned economy, the Ruble Crisis at the end of the 1990s, and a tragic airplane crash contributed to the closure of the coal mine at Pyramiden, and the settlement around the mine was abandoned in 1998.

As it stands today, Pyramiden is both a monument to a lost utopia and a testimonial to how industry creates the future. In this, the town is no different from industrial ruins all over the world. But Pyramiden in Svalbard makes an indelible impression, probably owing in part to its isolated position in the terrain, facing the spectacular Nordenskiöld glacier on the other side of the fjord. This town without inhabitants lingers on as a sculptural imprint of human civilization in the Arctic, forsaken by and reunited with its awesome natural surroundings.

Although production and coal mining under the mountain at Pyramiden have ceased, and the settlement lies abandoned, industrial society lives on within many of us in the form of customs, habits, values. Pyramiden in Svalbard is a monument to Soviet socialism, a reminder of solidarity as an ideal, and of industrial society as a lost utopia.

But this essay is not about *that* Pyramiden, not the well-ordered city, not the abandoned mine shafts in the mountain: it is about the mindset rooted in that inner, utopian Pyramiden we carry with us from industrial society and the century of the working class. This fossilized mindset is not particular to the Arctic, but it is in polar regions that the consequences of this mentality will appear first and most clearly.

Ore, as we know, is an economic term, not a geological one. In the case of Svalbard's coal, it is tempting to say it is a *geopolitical* rather than an economic term. Neither the Soviet Union nor Norway ran mining operations for the monetary returns; the payoff lay in Svalbard's strategic position during the Cold War. For both parties, coal mining was a costly pretext for being in an advanced position for eavesdropping. But once the coal was mined, it might as well be used for something. In 1961, Stortinget (the Norwegian Parliament) established the coke plant Norsk koksverk in Mo i Rana, to ensure sale of coal from Longyearbyen and a supply of coke to the ironworks at Mo. Coal from the mines in Svalbard competed with coal from other producers on the global market.

*

At the turn of the millennium, just a few years after the mine at Pyramiden was closed, I was traveling in India for a book project.[31] My journey began in Mumbai; from there I went to the capital of Delhi and on to Kolkata on the Bay of Bengal before turning south through Orissa toward the ancient town of Madras, which is now called Chennai. At some point, probably toward the end of my journey, the name Bhopal cropped up, reminding me of what must still be considered the greatest industrial disaster of all time.

The night between December 2nd and 3rd 1984, poisonous gas started leaking out of a factory that produced herbicides in that town. The factory was owned by the American Union Carbide Corporation; the gas, which contained methyl isocyanate among other things, spread quickly and invisibly to the people sleeping in the slums surrounding the factory. According to official statistics, about 4000 people died immediately from poisoning, 8000 died within the two following weeks, and another 8000 died later on as a result of the gas leak. These figures are probably a bare minimum. In 2006, an Indian government report estimated that over half a million people were injured, thousands of them seriously. Union Carbide has never acknowledged responsibility for the disaster; no one in management has been punished, and since the corporation has now become part of the Dow Chemical Company, it is even more difficult to take legal action against those responsible.

If I was especially interested in this case, it was for a particular reason. In its heyday, the Union Carbide Corporation was a multinational concern with mines, factories and other industrial plants all over the world. UCC had also established two companies in Norway. It ran a smelting plant in Meråker in Trøndelag in central Norway, but Union Carbide's biggest plant was the smeltery in Sauda in Ryfylke in southwestern Norway. This was a conversion plant, where manganese ores were melted and converted to alloys such as standard silico and ferromanganese. The ore arrived from places like Ghana in West Africa, Amapá in Brazil, Orissa in India, and from South Africa. Most of the coal was shipped in from British ports such as Newcastle and Cardiff, and was crucial for the smelting process. Coal was crucial, but at the same time, the objective was to *rid the metal of coal*, leaving as little carbon as possible in the end product.

This use of raw materials from around the world meant that production was *globalized* long before the word entered the Norwegian language. In addition to factories in India and Norway, Union Carbide had a head office in Manhattan in New York, a shell corporation in Canada (named Electric Furnace Products Company, Limited), mines all around the world, and an accounts office in Cragmore in Bermuda. The booty was buried on Treasure Island, so to speak.

The formal owner of the smelting plant in Sauda was the Canadian company EFP Co Ltd, which was essentially nothing more than a mailbox in Toronto. This smelting plant, this production, this international trade, and the social life centered around the factory formed the backdrop for my entire childhood. Nearly every man I knew spent his entire life working at EFP, as the factory was usually called. Like most of my peers, I took shifts as a factory worker every summer vacation after I turned 16, and worked even more after high school, a total of two or three years. When I started publishing books, the smelting plant was a central theme and the main source of metaphors in my writing. Electric Furnace Products Company, Limited, was an enterprise I knew inside and out, knew how its production was organized through the post-war period up until 1990, when the smelter celebrated its 75th anniversary. Socially, I am a true child of the industry, a son of Union Carbide, with industrial ethos as my most important moral compass.

During my childhood, pretty much all of the industrial district around the factory lay under a thick blanket of manganese fumes, rain and fog. Our boulevards were the pipelines, full of rushing water and pure energy, that plunged from the upland moors of Ryfylke to the power plants at the inner end of the fjord. The layout in Sauda was just the same as in Odda, Tyssedal, Ålvik, Svelgen, Øvre Årdal, Glomfjord. Standing beside the power plant cathedral, there was always a smelter spewing thick smoke into the already thick layer of clouds. And it rained without cease. On clear days the layer of smoke would be particularly dense; to put it drastically, I was twelve years old before my father took me up the mountain and showed me the sun for the first time. What a dazzling, powerful, almost unreal sight that was! Below us the smoke hovered like a thick, gray, impenetrable lid over the end of the fjord and the roofs of the town. As we descended under the edge of the smog, we sensed the sharp, distinctive smell of manganese, and thought what many said out loud: it smelled like money; it smelled like a steady job with a regulated salary; it smelled like progress and development.

After industrialism's blissful and (not least) blissfully *unaware* moments in the 1950s, ideas about environmental protection and reducing harmful emissions got through even to energy-intensive industries. During the 1960s, the manganese fumes were cleared from the air, leaving the factory looking like a ship run aground, no smoke coming out of its smokestacks. Yet it was more productive than ever before. Where had the smoke gone? At first the solution looked like a stroke of genius. With simple technical measures, the fumes from the factory chimneys were led into the sewers and flowed untreated into the fjord. I still remember how happy we were, how proud of this technical advance, as we stood down by the unloading dock and watched the thick brown sewage gush out under the quay. The smog was gone, the sun was shining; that's what progress and development looked like! Above us the sky was blue – and below us, the fjord was slowly being laid to waste, fish and seaweed disappearing. And the ethical aspects of this fact never came up for discussion; production was A Good Thing, any drawback was a necessary side-effect.

In the words of the Norwegian folklorist Svale Solheim:

> The labor of creating things is where our forefathers drew the dividing line between right and wrong, truth and falsehood, good and evil. Our collective sense of justice emanates from the manufacturing process itself and can be summed up briefly: Anything that served to advance creative labor and secure the best possible result – quantitatively and qualitatively – was right, and simultaneously took on the ethical quality good. It was right and good because it represented the life force; it sparked life and it sustained life.[32]

In addition, some types of labor have existential and heroic dimensions. Deep-sea fishermen and miners have always trodden the fine line between life and death. These workmen sacrificed themselves for our prosperity. George Orwell wrote, "...all of us *really* owe the comparative decency of our lives to poor drudges underground, blackened to the eyes, with their throats full of coal dust, driving their shovels forward with arms and belly muscles of steel."[33] For many years, the same words could be said about the miners of Svalbard, in Longyearbyen, Svea, Barentsburg. And the words are even more true of today's oil workers in the North Sea. By subjugating fire, the men who labored around the smelting plant furnace also approached this heroic status. Despite fumes and grime and poison and discharge to the environment, the furnace workers appeared as promethean heroes who had snatched fire from the gods and given it to humankind. In their hearts they felt they were part of something momentous, something greater than merely melting metal: they were also shaping a shiny bright metallic future.

In the mountains above the industrial towns of Scandinavia's rainy coast, most of the tarns and streams were dammed and funneled into pipes. The first few decades after production started, smoke lay thick over the factory towns, and later the fjords were laid to waste. In the smeltery, shift workers labored night and day; most jobs involved hardship, noise, smoke and fumes. When we inhaled smoke and metal dust, we said it was no worse than a proper workman should tolerate. Physical danger was also part of our workplace; explosions and falls claimed many lives. Despite all this, it is safe to say that everyone I knew considered their drudgery in industry a major step forward, a radical liberation compared to slaving under nature in the primitive farming we had left behind. For most of us, working in industry represented huge improvements in standard of living, quality of life, and freedom. Such thoughts were on the minds of many of the factories' women and men as they stepped out through the gates and into the post-industrial world.

Industrial manufacturing, with all its trappings, was the vital, creative labor that not only generated goods and services but, above all, created the future – a better future. "Conversely, anything that raised obstacles for creative labor and threatened to destroy what it accomplished was *wrong*, and took on the ethical quality *evil*. It was wrong and evil because it represented forces opposed to life; it threatened life itself, threatened continued existence."[34]

For a long time, many continued to view environmentalists and tree-huggers as representatives of forces opposed to life.

*

From the hills overlooking our plant, no other sign of civilization can be seen.

That is how the industrial town of Sauda was presented in Union Carbide's company newsletter sometime in the 1970s. We may smile at this innocent arrogance, but the sentence also reveals some truths. Industry in its classic form was Janus-faced, two-sided. On one side it stood for the great advance of civilization; on the other side, for barbarity. In Union Carbide's case, the barbaric face was revealed most blatantly in Bhopal, in the Indian state of Madhya Pradesh; the industry's civilizing face gazed out over the landscape around the factory in the little fjord town of Sauda in Ryfylke.

Smelting pure metal out of manganese ore requires raw materials – coal, coke, quartz and limestone – and copious amounts of electricity. This fact also made the smeltery part of a complex production system. Even though living conditions in Bhopal and Sauda were completely different, the living conditions and future of both places were dictated not by political authorities in Oslo or New Delhi, but in a board room in Manhattan in New York, by people none of the workers had met and who couldn't be made responsible. When I was a child, I didn't know anyone who had been to Oslo, because no roads led there. But everyone had been to a city called Takoradi in Ghana, because that was where the manganese ore came from, arriving on ships of the Norwegian–America Line about twice a month.

The electricity was generated locally from the rain and heavy snowfall over Scandinavia's west coast. Quartz and limestone came from small quarries along the Norwegian coast; coal and coke came, not from Svalbard, but from mines in Yorkshire and Wales. Union Carbide ran its own manganese mines in what was then called the Gold Coast, currently Ghana in West Africa; for many decades that was where most of the ore came from. When those mines had been emptied, Union Carbide had to look around for other sources of ore.

With ready access to copious amounts of hydroelectric power on the Rain Coast here in the North, workers transformed the cheap ore from the Gold Coast into manganese slabs, which ultimately became finished products in the big industrial nations. As part of an unjust post-colonial world economic system, this process in itself posed a moral dilemma. But for the smelting plant in Sauda, something else happened to make the problem acute. In the 1980s, the best, purest manganese ore came from South Africa. Management claimed that without South African ore, the company would not survive. The 1980s was also the decade when efforts to isolate the apartheid regime were at their peak. Strong voices demanded that South

African products be boycotted. By this time, the Norwegian metallurgical company Norske Elkem had taken over the smelting plant in Sauda from Union Carbide. Here let it be said that the Norwegians were the worst, most short-sighted owners in the history of the Sauda smeltery. In the decade between Union Carbide and the French concern Eramet, which currently owns the plant, the emphasis was more on cutting back production than on investing for the future. On top of that, Elkem faced the demand for a boycott. And this demand confronted not just Elkem's company management, but also the local union, Shop 31 of the Norwegian Union of Chemical Industry Workers, which had a long, red, and radical history.

Now management and the union faced another acute political dilemma with moral overtones. The choice stood between breaking the international boycott of South Africa and

Operations in Pyramiden ceased overnight in 1998 (Photo: Leif Magne Helgesen)

contributing to the demise not only of the plant, but of the entire community – at least if the company managers were correct in their belief that the South African ore was crucial to the plant's survival. Management and the union at the smelting plant struggled with this moral dilemma until it resolved itself when the apartheid regime in South Africa surrendered power to the majority and Nelson Mandela left his cell on Robben Island and descended onto the political stage like a *deus ex machina*.

*

Through pure coincidence (which occasionally resembles fate), my authorship is intertwined with industry and mining. In addition to single-employer industry towns like Sauda and Odda, which I know very well, I have lived and worked in the mining town Bjørnevatn in Sør-Varanger and the ore-shipping port town of Narvik, both north of the Arctic Circle. I have described them in essays and fiction. My book *Pyramiden*[35] from 2007 is about the abandoned Soviet coal mining town at the inner end of Isfjorden in Svalbard. I have visited and written about the mining towns of Norilsk in Siberia, and Nikel and Zapolyarny on the Kola Peninsula. In the southern antipodes, I have lived in and described mining towns like Lota and Coronel on the Chilean Pacific Coast, and the copper mines in Chuquicamata, which lies farther north, in the lunar landscape of the Calama desert at the foot of the Andes. In keeping with my background and mindset, I have felt it natural to describe the activities in all these places as heroic reaping of natural resources, for the benefit of the owners, but also for the common good. It isn't just that I speak the iron alloy dialect; I am myself a typical alloy of protestant work ethics and modern industrial ethos. This mindset is hard on the surface and resilient inside, like high-alloy steel. The industrial way of thinking is strong and conservative and resistant to change.

Religious aspects have never been central to my world view – neither in my life nor in my work. Nonetheless I observe that as soon as I go south of Drøbak, I am a Lutheran protestant, as good a representative of Christian culture as any. The protestant faith, as we know, strongly embraces the principle that labor – especially hard manual labor – is right, and also assigns it a moral dimension: work is good in and of itself. Much of this mindset lived on in the socialist tradition. Now, increasingly, the morality of this way of thinking is challenged by growing insight into the harmful effects of mining and heavy industry on humans and the environment.

At this time in history, it is incontrovertible that not only extracting raw materials from the natural environment, but also refining those raw materials to semi-finished and finished products, presents a moral dilemma. Honest work is no longer wholly right and good, as it was in the protestant and socialist world view. On the other hand, if we stop producing goods in the affluent world and move the factories to poor countries, we risk more than simply causing even more dramatic damage to the environment than we do today. When ore is smelted and consumer products are assembled in Indonesian, Brazilian and Chinese sweatshops, it is done under minimal demands for environmental safety and low emissions. Working conditions are similar to those we saw in Europe at the beginning of the industrial revolution, devoid of the rights the European working class has struggled for centuries to achieve.

"The light work sheds is a beautiful light, which, however, only shines with real beauty if it is illuminated by yet another light," wrote the philosopher Ludwig Wittgenstein.[36] Another tradition maintains that work – as the intermediary of exchange between man and nature – is the living, creating fire. We now know that both retrieving resources from the earth and refining them (if we can still call it refining in this context) have dark sides.

Raw materials and production is one thing. What the end products are used for is another thing entirely. Clearly, the industrialization of places like Sauda and Odda and Longyearbyen and Pyramiden conferred the advantages of civilization on the towns that grew up around the mines and factories. Worker involvement in management, strong unions and widespread union membership helped guarantee that.

Union Carbide's plant manufactured semi-finished products – various types of manganese alloys. Manganese is used as an additive in the production of steel; it makes steel hard on the surface and resilient inside. This is high-alloy steel. Manganese steel is top quality steel, used for precision instruments, engine parts, and – not least – weapons. When the semi-finished manganese slabs from EFP in Sauda were loaded onto cargo vessels, they usually crossed the North Sea to ports serving the German Ruhr district or the English Midlands. From the docks in ports such as Duisburg in Germany and Scunthorpe in England, the manganese was transported to steel mills and armament factories.

The reconstruction after World War II meant a golden age. Not only war and peace, but also *rumors* of war and peace were easy to detect in the community in the form of higher tax revenue and more money for municipal services. Acts of war and international crises showed up locally as more day-care centers, new schools, better care for the elderly. Here, clearly, one man's meat was the other man's poison.

In the poison disaster in Bhopal, industrial capitalism showed its most barbaric and repulsive face. Industrialism has also been a civilizing, community-building force. The life quality that many of us now enjoy would be unattainable without industrial production on a grand scale. When factories transform natural resources into semi-finished goods that go into deadly weapons, they also create zones of new life, prosperity, and civilization. This has been – and remains – the fundamental moral dilemma of heavy industry. In Sauda, the fumes that polluted the air and the poisons that spewed into the fjord were eventually captured and subsumed into the production process. As industrial society gradually stretches beyond the limits of what nature can tolerate, it looks as though the advanced weaponry wrought from manganese steel should not be aimed at alien forces, but at the industries themselves. That would solve several practical problems and perhaps a few moral problems as well. But the basic dilemma remains.

In *Ars Poetica* (The Art of Poetry) from 18 B.C., the Roman poet Horace warned authors not to resort to invoking what he called a *deus ex machina* (god from the machine). In the Greek tragedies Horace was referring to, dramatic impasses and impossible situations would often be resolved by winching an actor playing the part of God down onto the stage. Horace condemned the use of *deus ex machina* as contemptible and false poetry.

Through work, humans have always used nature's riches to break the bonds of slavery nature imposed on us. Now we are enslaved by the machines we created to set ourselves free. This is a dilemma we must solve on our own. In the fateful drama of our environmental politics, we cannot rely on gods descending onto the stage. The light work shed was a beautiful light; it has created progress, freedom and civilization, but a civilization that now casts long shadows and has left behind poisonous landfills, slag heaps and mountains of broken rock. Coal and coke burn with a pale blue flame. And a blue warning light is burning for carbon-based industry. To shine with real beauty, the light work sheds must be illuminated by yet another light – one we may be able to discern at the end of the long industrial thought process that began in our inner, utopian Pyramiden.

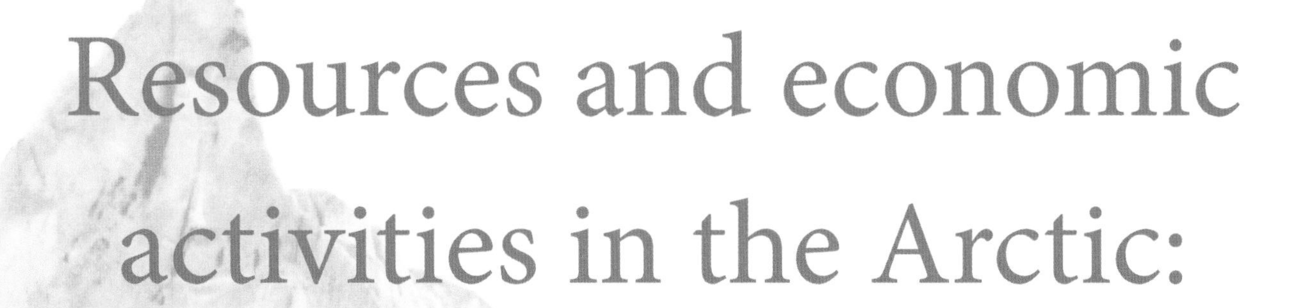

Resources and economic activities in the Arctic:

shipping – coal – fishing – oil and gas

Introduction

by Gunnar Sander, Senior advisor, Norwegian Polar Institute / PhD Candidate, UiT – the Arctic University of Norway

The interest in new commercial activities in the Arctic has grown noticeably in the past few years. Discussions in Norway and other countries show clearly that the petroleum industry wishes to expand northward. But oil and gas are not the only energy resources; the Arctic also contains both coal and novel energy sources such as gas hydrates, and there is obvious potential for renewable energy from water, wind, and waves. While Norway has focused mainly on oil and gas, several other arctic countries have been more interested in developing mining. Mining has been going on for many years in Svalbard and parts of Russia, Canada, and Alaska. New deposits of ores and minerals have been found, along with precious metals and diamonds. Given that huge areas of the Arctic are still relatively unexplored, there is good reason to believe that even more new resources will come to light. The region also contains biological resources such as fish, shellfish, marine mammals, game, and forests. Species that have evolved to survive arctic cold and darkness may hold the keys to new products in the growing biotechnology industry.

This growing interest is often presented as a consequence of climate change. Warmer temperatures will obviously melt away sea ice and glaciers, making the resources more readily available. But many overlook the other side of the coin: melting of rivers and lakes, and loss of permafrost reduce accessibility on land in many places. Where it was once possible to drive over frozen surfaces, new transportation infrastructure must be built. Construction on the melting permafrost necessitates extensive pile-driving and excavation, raising costs. Climate change can thus make conditions for commercial activity in the Arctic both better and worse.

However, the driving forces behind new economic activities are not to be found in the climate system. Growth in commercial enterprise is often vaguely attributed to "globalization". Against a backdrop of economic expansion and population growth, some sketch an image of a race for the resources in the north. But the fact that those resources lie in the Arctic does not guarantee that they will be exploited. It is costly to work under stern arctic conditions, and the customers are far away. If it is cheaper to buy the same raw materials from elsewhere, there is no market for what the Arctic has to offer. Vast forests remain uncut, and it has become even more difficult to develop arctic gas resources after extraction of shale gas intensified and gas prices came down: these are examples of how the market rules. A desire to ensure access to strategic resources and reliable supply from peaceful regions is a consideration that may counterbalance the disadvantages and favor development in the Arctic, despite the costs. Costs may also increase due to environmental regulations. The parts of the energy sector that rely on fossil fuel can envisage scenarios where both the climate and market forces might put a stop to further development. How these factors will affect various commercial sectors is difficult to predict. Piet Hein's classic aphorism appears to be true for much of the economic development in the Arctic: *Things Take Time.*

This part of the book gives examples of three industries in the Arctic: petroleum, fisheries, and shipping. The three chapters discuss factors that may influence how commercial activities develop. In the last section, some of the ethical aspects of development will be presented.

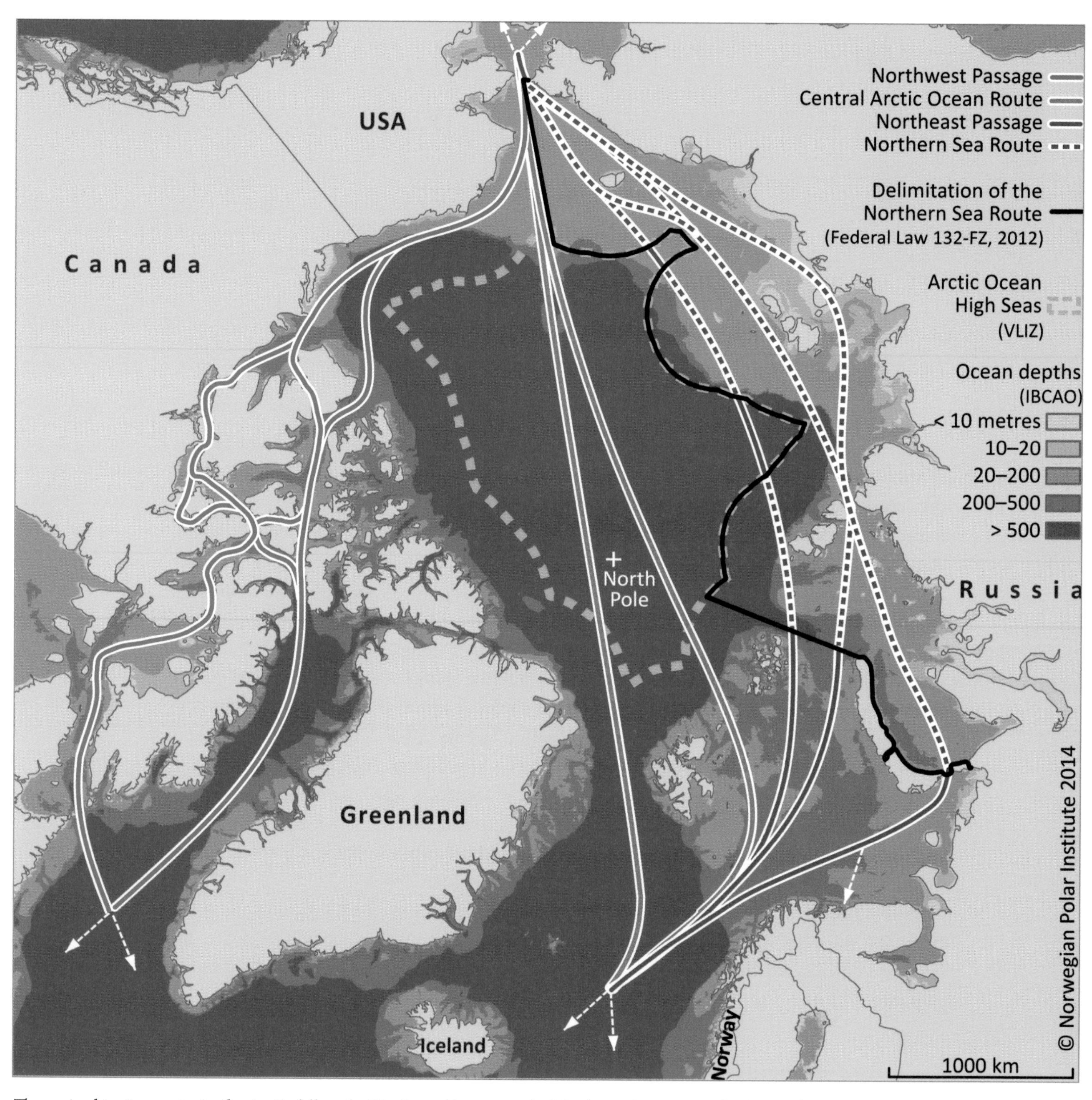

The main shipping routes in the Arctic follow the Northeast Passage or the Northwest Passage. In the future they may also go straight across the Arctic Basin. Note that the Northern Sea Route is the Russian part of the Northeast Passage (Map: Norwegian Polar Institute)

Shipping in the Arctic Ocean

by Gunnar Sander, Senior advisor, Norwegian Polar Institute / PhD Candidate, UiT – the Arctic University of Norway

From newspapers, television studios and political podiums, we often get modern versions of the dream that once lured Willem Barents to the shores of Svalbard: the hope for new trade routes linking east and west. How likely is such a route today? What impact might it have, and what can we do to meet any challenges?

Trade routes and traffic

The maritime history of the Arctic Ocean abounds with tales of ships that have struggled against the ice, often losing the battle. Nonetheless, there has been fairly extensive traffic in Russian waters. After the revolution and the intervention wars, the new Soviet government strove to expand northward and industrialize Siberia. Transportation infrastructure was built up around the vital arteries of the trans-Siberian railway, the major navigable rivers, and along the northern coast. Traffic along the Northern Sea Route (see map) peaked in 1987. But during the economic restructuring that ensued after the introduction of the market economy in the 1990s, the fundament fell apart. Industries and ports closed down, the population fell, and the infrastructure disintegrated. Despite substantial increases in traffic in recent years, shipping activity in 2013 remained just two thirds of what it was in 1987. Preliminary statistics for 2014 indicate a significant drop in activity.

On the other side of the Arctic Basin is the route we often call the Northwest Passage: several alternative pathways through the archipelago of northern Canada. Scattered communities have supplies shipped in, and a growing number of pleasure vessels pass by, but there is little other activity that creates maritime traffic. In the central Arctic Basin, traffic is still negligible, apart from occasional research vessels and a few cruise ships taking tourists to see the North Pole.

What can we expect?

Russia has ambitions to develop the Northern Sea Route. The country is investing in ports; centers for search and rescue have been designated, and a program has been adopted to replace the ageing fleet of nuclear-powered icebreakers. As before, one of the central objectives is to contribute to economic growth in northern Russia – a region rich in oil and gas fields, minerals, and forests. Better transportation is a prerequisite for making use of these resources. For example, exploitation of the enormous gas fields on the Yamal Peninsula requires that materials and huge prefabricated structures be shipped in. Once the operation is running, the industry itself and the surrounding communities will need supplies, and the gas that is produced will be compressed into liquefied natural gas and shipped out.

Natural resources are expected to be developed in other arctic countries as well: new mines, oil and gas discoveries, and perhaps new fisheries. The common denominator for all this is commercial activity *within* the region – be it on land or at sea. Such activities explain most of the increase in shipping traffic over the past few years, including that along Russia's Northern Sea Route. Regional activities are also expected to be the most important driver of increased shipping in the Arctic Ocean in years to come. But just how much, where, and how rapidly traffic will increase remains to be seen. That depends on a number of uncertain factors that influence economic developments, such as market prices of raw materials and development of new technology for arctic conditions.

What about intercontinental traffic? So far, it is only just beginning. In 2013, eighteen ships sailed all the way through the Northeast Passage. The same year, the first cargo ship in transit sailed through the Northwest Passage. While the Russians clearly want the international fleet to be able to pass through their waters in transit, and are doing what they can to make it possible, the Canadians appear more reluctant.

Several obstacles make significant transit traffic in the Arctic Ocean unlikely in the near future. First, the market is limited. It is only routes between ports in northern Asia and northern Europe that are shorter than routes through the Suez Canal. For ports farther south, the northern route is a detour. Second, sailing in the north it is still too unpredictable because of the combination of harsh ice and weather conditions, and inadequate infrastructure and support services. This makes the route poorly suited for the largest segment of international transport, namely container vessels that are a link in a transport chain that does not tolerate delays. In addition, ice conditions limit the sailing season to just a few months per year. This means that regular transports will require different logistics summer and winter, an unattractive prospect for shipping firms – at least so far. Finally, several factors related to sea ice make it more expensive to sail northward than southward (ice-classed vessels, icebreakers, fuel consumption). Only when sailing across the Arctic Ocean can compete in terms of price, reliability, and safety, will the intercontinental stream of goods choose this route. Until then, they will follow other shipping routes (Suez and Panama Canals) or other means of transport (railways and pipelines).

Impacts of shipping in the Arctic

Economic and societal impacts

Arctic shipping is both a prerequisite for and a consequence of economic development. When infrastructure is expanded to support arctic shipping, one of the most important ramifications is how it contributes to exploitation of natural resources in the region. Construction of new harbors, upgrading of existing harbors, and establishment of emergency supply stores and monitoring stations will have direct economic effects. Indirect effects will arise through delivery of goods and services for shipping, such as fuel, supplies, and repairs.

Which individuals will find employment and enjoy a share of the income will depend on which qualifications are sought, and what skills the local workforce has to offer. Many small arctic communities have little in the way of specialist competence. Major expansion in such places will entail influx of workers, possibly as commuters, until local residents have eventually been trained for the jobs. While the local people who get new jobs may enjoy higher income, better housing and improved access to goods and services, the gap to those who do not benefit from the change will grow. This is one of several possible roots for social problems.

There is reason to assume that economic activities will be unevenly distributed. Major mining operations, petroleum fields, and industries will obviously generate extensive transportation activities. Port towns such as Murmansk, Churchill, and perhaps Kirkenes may evolve as hubs for transfer of goods. But activity in most harbors in the Arctic will probably be low, particularly since ships in transit have no reason to stop at harbors along the way – unless they are forced to.

Outside the Arctic, increased supply of raw materials will mean stiffer competition, with a greater variety of goods and lower prices. The Arctic can also contribute toward increasing the security of supplies to Europe and the United States. However, the political tensions and economic sanctions in the wake of the crisis in the Ukraine cast shadows on the collaboration between Russia and the West also in the Arctic. There is hope that intercontinental transit traffic will contribute to bringing down transportation costs. Trade between regions with favorable locations may increase. That could spark growth in the economies involved. Maritime traffic in the Arctic will also create new markets for shipowners, shipbuilders that make ice-strengthened vessels, and maritime services. Norway is in a good position to benefit from this, but industries in several EU countries and Asian countries such as South Korea and Japan are well placed to do the same.

Safety and the environment

It is riskier sailing in the Arctic Ocean than in other seas. Threats include sea ice, icebergs and growlers, vessel instability owing to icing on the superstructure, polar low-pressure systems, low temperatures and polar night. The seas are inadequately charted and marked, communication devices function poorly near the poles, systems for traffic monitoring are sketchy, and few mariners have adequate qualifications for these waters. Taken together, these factors mean an increased likelihood of accidents and shipwrecks. The chances of getting assistance in emergencies are limited, as search and rescue resources are underdimensioned for these vast expanses of ocean. Recognizing this, the Russians send icebreakers to accompany ships on the Northern Sea Route. Cruise ships can reduce risks by sailing in convoy, though this strategy has not been widely used by operators in the Arctic. Large cruise vessels represent the greatest risk for loss of human life. The biggest cruise ships that sail to Svalbard take 3300 passengers; if an evacuation were ever necessary, it would stretch rescue resources to the limit.

Incidents and accidents can cause all types of cargo to end up in the sea – including cargo that is poisonous or in other ways harmful to the environment. Oil spills are feared most. After an accident, it is fairly common that the ship's own fuel, bunker oil, leaks out. How much damage this causes depends on whether the ship runs on heavy fuel oil or lighter fuel, such as diesel. Accidents with tankers can cause truly massive oil spills, like the one after Exxon Valdez ran aground in Alaska in 1989, but are fortunately rare. The Russians run many tankers in their northwestern waters, with a modern fleet that also ships crude oil along the coast of Norway. Along the Northern Sea Route, the tankers have so far mainly transported lighter types of oil, such as aviation fuel and liquefied natural gas. If crude oil from the Pechora region were to be shipped eastward toward Asia, the risk of serious damage would increase. Given currently available resources, the possibilities for cleaning up oil spills in the Arctic Ocean are small, particularly in ice-covered waters. The possibilities would be equally small after spills of chemicals or radioactive substances.

Ships may legally discharge rubbish, sewage and oily waste from machine room and tanks, provided the substances are diluted and discharged far enough away from land. Emissions to the atmosphere usually contain ordinary combustion products such as sulfur and nitrogen compounds, CO_2 and soot. In winter, the polar atmosphere is cold, dry, dark, and stratified. Emissions at this time of year can remain in the stable air masses until spring, when new processes come into play. Air pollution from ships causes considerable damage to health globally, but in the sparsely populated Arctic, few people are exposed. The air pollution also causes acid rain and harms vegetation. The effects on climate depend on where the emissions occur. In particular, soot and some short-lived greenhouse gases mean that emissions in the Arctic have greater effect on climate than the same emissions would have farther south. The paradoxical result is that moving shipping from the Suez Canal to the

Arctic Ocean may *increase* global warming for more than a century, despite shorter shipping routes and lower total emissions.

Ships are the most important vector for transmission of non-native species to new parts of the oceans. Algae, jellyfish, shellfish and crabs are among the species that have been found in ballast water tanks and on ship's hulls. If the stowaways survive the trip, are released and do well in competition with local species, there can be major changes in the marine ecosystem and significant losses in monetary terms. So far, the difference in temperature between Svalbard and seas farther south has made it unlikely that non-native species would survive the trip north. With rising water temperatures in northern seas, this barrier may be more easily overcome. However, shipping east–west, as in the Northeast Passage, essentially follows temperature isolines. These routes also pose a greater risk because there are more species in the northern Pacific Ocean than in the Atlantic, where non-native species may find empty niches to fill. The king crab, also known as the Kamchatka crab, and the snow crab are examples of species that originate from other northern waters and that now thrive in the Barents Sea.

What can be done?

A two-pronged strategy is generally recommended to meet the challenges of shipping in the Arctic: improving international regulations and building up a maritime infrastructure suited for arctic conditions.

Shipping is an international industry, and is primarily regulated through global institutions, mainly the International Maritime Organization (IMO), an agency under the United Nations. IMO has an extensive set of regulations including about 50 conventions and a series of recommendations and technical standards. All of these are global, and not particularly suited to the challenges in the Arctic. Negotiations are therefore currently ongoing to establish a mandatory Polar Code, which will probably be open for signature in 2015. The goal is to establish regional shipping regulations that go beyond global regulations, so that the risks to safety and the environment are acceptable. So far, there have been few analyses of the specific standards under discussion, and who intends to back which proposals. Scandinavian shipping organizations have generally supported high ambitions, but voices have also expressed concern at the prospect that some requirements may in effect stop traffic in the Arctic before it even gets started. Environmental organizations have expressed concern that the code does not cover all environmental issues in the Arctic.

However, global forums do not make all the decisions. The law of the sea gives coastal states broader jurisdiction to regulate navigation and require certain standards of vessels operating in internal waters and the territorial sea. In ice-covered waters, coastal states can also set standards aimed at protecting the environment out to 200 nautical miles. Canada and Russia have used this to create extensive national regulations, which foreign vessels are obliged to follow. This is controversial. It is likely that the two countries will face demands to harmonize their regulations with the Polar Code, once it has been adopted. This illustrates how the law of the sea balances the rights of coastal nations to protect themselves against the threats posed by shipping, and the shipping nations' wish that their vessels be able to sail as freely as possible, without special regulations from other countries. If the Polar Code and other IMO regulations are insufficient from the perspective of the Arctic Ocean coastal states, coordination of their national regulations is an interesting option for eliminating the weaknesses, while ensuring that ship owners meet the same standards in all arctic waters.

Development of infrastructure and support for shipping will predominantly fall to arctic coastal states. They have signed agreements for search and rescue operations and oil-spill preparedness and clean-up in the Arctic Ocean, to help each other out in case of accidents. So far, they cannot fill these agreements with satisfactory content owing to a lack of equipment and infrastructure. The coastal states also share the crucial tasks of charting the seabed, marking shipping lanes, monitoring shipping, and enhancing support systems in ports. Other countries can contribute, for example by providing better navigation and monitoring via satellite, technological innovation, and better forecasting of weather and ice conditions. Their ships can bring income to the coastal states in the form of harbor fees and fees for use of services, such as those Russia charges for icebreaker escorts. This type of income will probably be small compared to the cost of development. The question is, then, whether the coastal states are willing to facilitate, and able to pay what it costs to ensure safe, environmentally responsible shipping in the Arctic.

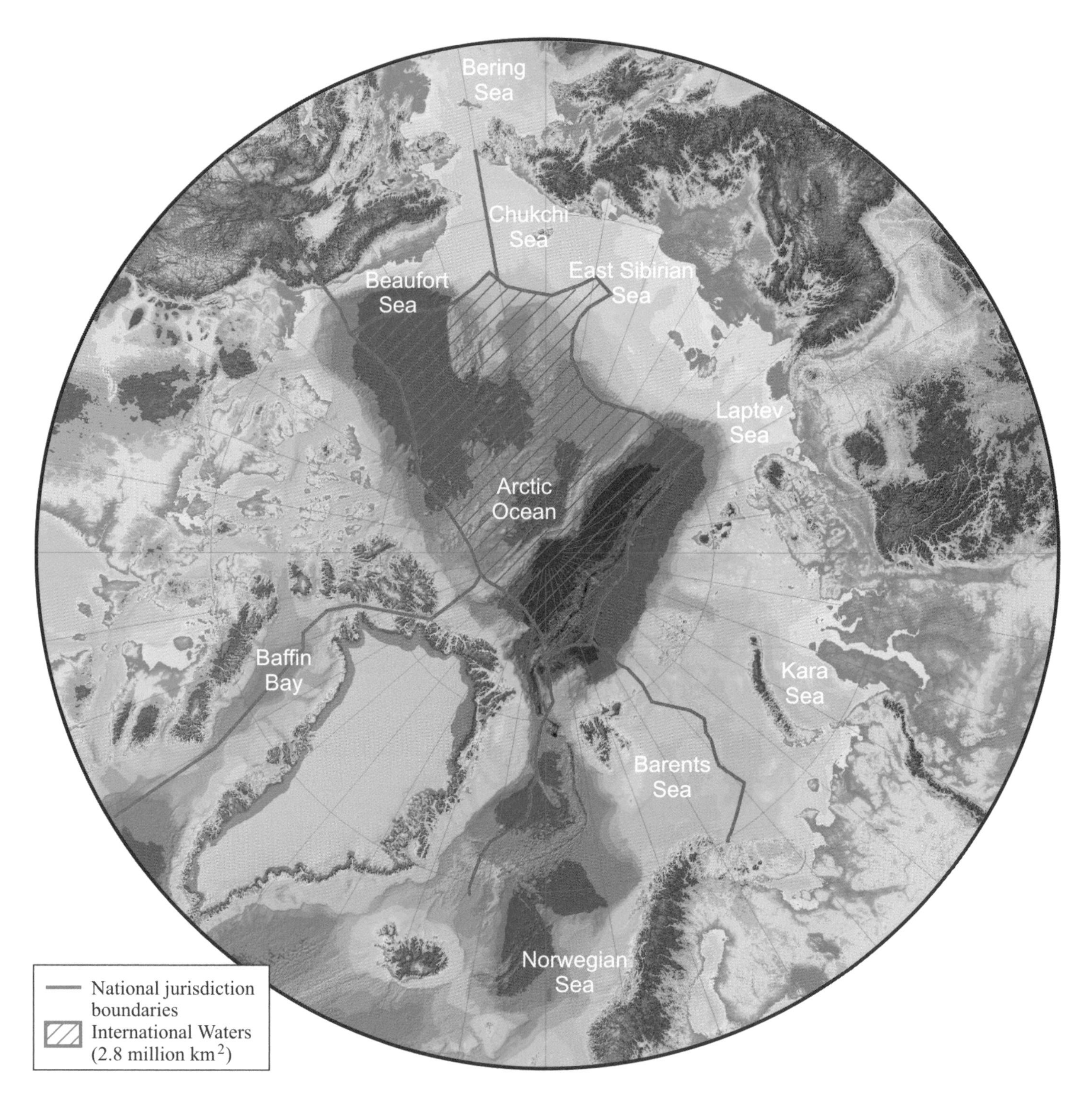

The Arctic Ocean. International waters and the 200-mile zones of the Arctic coastal nations are indicated (Map: Institute of Marine Research)

Fish and fisheries management in the Arctic Basin

by Alf Håkon Hoel, Research Director, Institute of Marine Research

The Arctic Ocean lies north of the land masses in the Arctic, surrounded by the coastal states of Norway, Russia, the United States, Canada, and Denmark/Greenland. In the center of the Arctic Ocean is a 2.8 million km² area of international waters, outside the 200 nautical mile economic zones under the jurisdiction of the coastal states. This is a deep ocean, covered with ice much of the year, and there is no fishing here. In the areas of the Polar Basin that fall under national jurisdiction, fishing is also negligible.

Fisheries in the north

In the coastal waters around the Arctic Basin, however, there is extensive fishing. Some of the world's richest fisheries are in the Bering and Barents seas, and in the latter they extend north toward the Arctic Ocean. The cod fishery in the Barents Sea and the pollock fishery in the Bering Sea are among the world's largest, each totaling about a million tons per year. There are also significant fisheries around Iceland and Greenland, and in the northwest Atlantic between Greenland and Canada.

Large populations of fish require vast feeding grounds, and warming waters have enabled species such as capelin and cod to move farther north. This has led to speculation that commercially attractive fish populations might move into the Arctic Ocean itself, and that the fisheries might follow.

If a fish population is to migrate, the new area must offer food, suitable water temperatures, satisfactory seabed conditions, and reasonable access to spawning grounds in shallow water. Given these criteria, it is unlikely that bottom-dwelling fish like cod and haddock will spread north into the deep ocean. Species such as arctic cod (*Boreogadus saida*), which live freely in the water column, might spread.

Fisheries in the Arctic Ocean?

Some worry that fishing vessels from distant fishing countries might take up unregulated fishing in the parts of the Arctic Ocean outside the 200-mile zones. Parts of this area have been ice-free in summer in recent years. In particular, an area in the Chukchi Sea north of Alaska and eastern Siberia has been of interest.

Speculation about fisheries outside the 200-mile zones in the Arctic Ocean has also prompted discussion about a possible fishery regime in the area. Several conferences have focused on this issue, and it has also been broached in conjunction with other meetings, such as those of the Committee on Fisheries (COFI), a body under the Food and Agriculture Organization of the UN.

The five countries bordering on the Arctic Ocean have been discussing these issues for several years. The basic premise has been that there are already extensive international agreements for the seas, including areas outside all national jurisdiction, and that these (obviously) also cover the Arctic Ocean. The Convention on the Law of the Sea from 1982 and United Nations Fish Stocks Agreement from 1995 require countries to cooperate in managing living marine resources outside the 200-mile zone. In 2008, the five coastal states signed the Ilulissat Declaration, where they reconfirm that they support the global conventions and will resolve any conflicting boundary claims peacefully.

In the North Atlantic, there are regional frameworks for fisheries management outside national jurisdiction, such as the Northeast Atlantic Fisheries Commission (NEAFC) and the North Atlantic Marine Mammal Commission. NEAFC has a mandate reaching all the way to the North Pole. The International Council for the Exploration of the Seas (ICES) provides assessments of the condition of living marine resources in the North Atlantic based on data and work contributed by marine research institutes in its member states; with this information as a foundation, ICES advises member states and regional organizations on the management of fish populations and the marine environment.

A series of meetings

In 2010, Norway organized a meeting in Oslo among government officials from the five coastal states to discuss issues raised by potential future fishing in the central Arctic Basin. The coastal states asked their research institutes to assess the situation. The following year, an international research conference in Anchorage concluded that commercial fisheries were unlikely to spread into the Arctic Ocean in the near future. Another conclusion was that additional research and monitoring are required for more careful analysis of these issues. Among other things, a paucity of monitoring data, particularly from areas north of Alaska, eastern Siberia, and Canada, makes it difficult to draw firm conclusions. Data coverage is better in the areas north of Norway and northwestern Russia.

Another meeting between the coastal states in Washington D.C. in 2013 asked the scientists for additional clarification concerning the likelihood that commercial fisheries would begin operating in areas outside the 200-mile zones in the Arctic Ocean. This meeting also discussed how unregulated fishing in the ocean might be avoided, and other factors of relevance for fisheries management in the area.

In October 2013, another research meeting was held in Tromsø. This conference took up existing monitoring of living marine resources in the Arctic Ocean and adjacent seas, and made recommendations for how it might be reinforced. As at the earlier meeting in Anchorage, they concluded that bottom-dwelling fish species such as cod and haddock are unlikely to find their way to the central Arctic Basin. This is a deep ocean, and even if the surface should be ice-free for longer periods in the future, the water is too deep for the groundfish to survive there.

On the other hand, pelagic species that thrive in cold water, particularly arctic cod (*Boreogadus saida*) could migrate northward if the ice cover shrinks and conditions are otherwise suitable. The circumpolar arctic cod lives over all shelf areas in the Arctic Basin, but is rarely harvested commercially. The only significant harvest has been by Russian pelagic trawlers mainly in the autumn near Novaya Zemlya, and harvest has been modest in recent years. In 2013, the stock was measured at under 500,000 tons.[37]

Fisheries management without fish

At a meeting in Nuuk, February 24–26, 2014, the five coastal states agreed on the elements of a draft declaration stipulating that, as an interim measure, they would not fish in international waters in the Arctic Ocean until a management plan is in place. This has long been Norway's attitude. Since 2007, Norwegian law has forbidden vessels sailing under Norwegian flag to fish in areas outside Norwegian jurisdiction unless there is a management plan backed by a regional fisheries management organization or something similar.

The Nuuk declaration also states that the five countries will continue with collaborative research on the resources in the Arctic Ocean, including a joint research and monitoring program. The next research meeting is planned for 2015.

Lastly, the declaration states that additional countries will later be included in a process of negotiating a more comprehensive agreement covering areas outside the 200-mile zones in the central Arctic Ocean.

Where in the Arctic Ocean might there be fishing in the future?

Even if the amount of ice in the Arctic Ocean continues to decline, most future fishing in the ocean is likely to be within the coastal zones of the coastal states, not outside national jurisdiction. The five coastal states all have large fishing fleets and extensive management plans for their waters. These management plans also regulate fishing in the Arctic Ocean, out to the 200-mile limit.

In places where commercially important fish stocks cross national boundaries, countries already cooperate within fisheries management; an example is the Joint Norwegian–Russian

Catch being transferred between Russian fishing vessels off Bjørnøya (Bear Island) (Photo: Leif Magne Helgesen)

Fisheries Commission. For nearly forty years the two countries have cooperated to manage the Barents Sea fish stocks they share (cod, haddock, capelin and Greenland halibut), and with time these efforts have paid off. The Norwegian–Russian joint management relies on extensive research collaboration between Norway's Institute of Marine Research and Russia's PINRO – the Knipovich Polar Research Institute of Marine Fisheries and Oceanography. The research collaboration also has a firm foundation in the International Council for Exploration of the Seas (ICES), which gives scientific advice to the Fisheries Commission.

If warmer water masses and less ice lead to growing fish stocks and a northward shift of fisheries into the Arctic Ocean itself, fishing is likely to continue mainly in areas inside the 200-mile limit. These shifts would become obvious long before there is any fishing in the Arctic Ocean itself, and they are also likely to affect commercially important fisheries. Norway and Russia already have a well-functioning collaboration for management of shared fish stocks, but such collaboration is not as well developed everywhere in the Arctic Ocean. For example, there is no joint American–Canadian fishery regime in the Beaufort Sea, since both fish stocks and fishing in the region have been negligible until now.

The big picture …

From a broader perspective, the most important thing about the developments described here is that the five coastal states are showing themselves and others that they are serious about their responsibilities as coastal states, and are at the forefront in meeting future challenges.

In areas where fish stocks are already large enough to support commercial fishing, management arrangements have already been established. At present, this means the marginal seas around the Arctic Ocean, such as the Bering and Barents seas. Less ice and warmer water may allow existing fisheries to shift northward to the shelf areas in the Arctic Ocean, but they will still be within the 200-mile zones. Significant harvest of fish beyond those zones, in waters over the Arctic Basin, is unlikely. A comprehensive global regulatory framework is already in place for how fish stocks within and outside the 200-mile zones should be managed. What the discussion between coastal states has added, is an obligation for these five countries to prevent their own vessels from fishing in areas of the Arctic Ocean that lie outside national jurisdiction, intensified research and monitoring, and a process to involve additional countries in these efforts.

Oil and gas resources in the Arctic

by Snorre Olaussen, Professor of Geology, University Centre in Svalbard (UNIS)

The world is currently dependent on fossil energy, and will probably remain so for several decades.

Owing to growing populations and demands for a higher standard of living in developing countries, the world's energy needs will increase dramatically in the next few decades. This is a reality we cannot escape.

It is a fact that over 1.3 billion people currently lack modern energy alternatives such as electricity. The world's total population, at present nearly seven billion, is expected to rise to 9.3 billion by 2050. It must also be acknowledged that the citizens of many third world countries use only a fraction as much energy as people in Norway and other western countries. For the anticipated economic development in many of these countries, more energy will be an absolute requirement.

The central question is therefore how we can get hold of enough energy to cover the expected increase. So far, despite increased investment in alternative energy sources, neither hydroelectric nor nuclear power has been able to replace fossil energy. Use of coal, oil and natural gas has remained stable, providing about 80% of all primary energy consumption. This percentage has not yet shifted in favor of alternative energy sources such as hydroelectric and nuclear power. Renewable energy sources, including biomass, have not contributed more than a tiny fraction of total energy consumption. According to British Petroleum's statistical report from June 2014, use of renewable energy sources is expected to increase from 2% today to 7% in 2035. If there are technological advances – which nearly always come unexpectedly – the picture could change rapidly, but so far, the world mainly relies on fossil energy.

In this context, the IPCC report, cautioning that two thirds of the world's known petroleum reserves must remain untouched if we are to avoid significant global warming, poses a formidable challenge. Even though many countries are implementing new political measures in the belief that this will limit the global temperature increase to less than 2°C, the world is drifting ever farther away from this goal – for local, practical, and economic reasons.

In the face of this as yet unresolved dilemma, all nations with oil and gas reserves have launched major efforts to chart their own petroleum resources. Some of the largest countries and energy corporations also want to have the best possible overview of the total global petroleum reserves and resources. Resources in the Arctic are no exception.

Part of this picture is that all five coastal arctic countries – Canada, Denmark/Greenland, Norway, Russia, and the United States – are charting their territories for potential future sources of energy.

There are currently over 400 active oil and gas fields in the Arctic, but they are concentrated to a few regions – Alaska, Pechora, and western Siberia. It is only in the last decades that the entire Arctic has been charted for potential oil and gas resources. This means that the estimates of conventional oil and gas resources in large parts of the Arctic, including the Norwegian part of the northern Barents Sea, are fraught with uncertainty; new and better estimates are needed. This is one of the purposes of this chapter; another is to discuss the possibility of more optimistic estimates of recoverable petroleum resources in the Arctic.

Terms and definitions

Petroleum is a broad designation for a mixture of hydrocarbons and other molecules, in solid form (for example asphalt), liquid form (oil) or gaseous form. *Fossil* energy sources are coal, oil and natural gas.

It must be clarified that unconventional forms of petroleum, such as oil and gas extracted from coal, oil sand, shale gas or oil, and gas hydrates, are not included. It is likely that both oil sand (tar oil) and shale oil and gas will be classified as conventional oil and gas in the near future.

This chapter is not intended to give a quantitative analysis of conventional hydrocarbon resources in the Arctic. For quantitative resource estimates, the reader is referred to the relevant institutions in each individual country.

Before we delve into estimates of petroleum resources in the Arctic, we must first define the area geographically. The United States Geological Survey (USGS) has chosen to define the Arctic as everything north of the Arctic Circle (i.e. north of 66°33'42.5''N, see figure on page 118), for the purpose of estimating resources.

In estimates of hydrocarbon volumes, USGS uses the unit "barrel" for oil and "cubic feet" for gas, whereas Norway uses standard cubic meters. This overview uses million standard cubic meters (10^6 Sm^3) for oil volumes and billion cubic meters (10^9 Sm^3) for gas volumes. The concept *oil equivalent* denotes gas recalculated as oil: 1000 m^3 of gas corresponds to 1 m^3 of oil.

Petroleum systems

Conventional petroleum resources are found in sedimentary basins. For such a basin to be able to contain conventional oil and gas, there are four geological prerequisites: a source rock, a reservoir rock, a cap rock (or seal), and traps that can hold the buoyant gas or oil. If all these formations are predicted to have been present at the proper time and space for formation and accumulation of hydrocarbons, geologists can assess the likelihood of finding an efficient petroleum system in the basin. Once this is clear, play models can be constructed and used for quantification of potential recoverable resources in that specific basin. Expected volumes are then presented in low, mean, and high cases.

Resources versus reserves

Unfortunately, there is some confusion concerning the meaning of the concepts *resources* and *reserves* among politicians, NGOs and media. This leads to misconceptions when an area's petroleum potential is being discussed. Fishermen in the north and west appear to understand these concepts better than Norwegian media and policy-makers (politicians). Simply put, a fisherman distinguishes between fish he has in his nets and can sell (reserves) and fish swimming free in the ocean (resources). In the oil industry, economic considerations are also used to distinguish between resources and reserves.

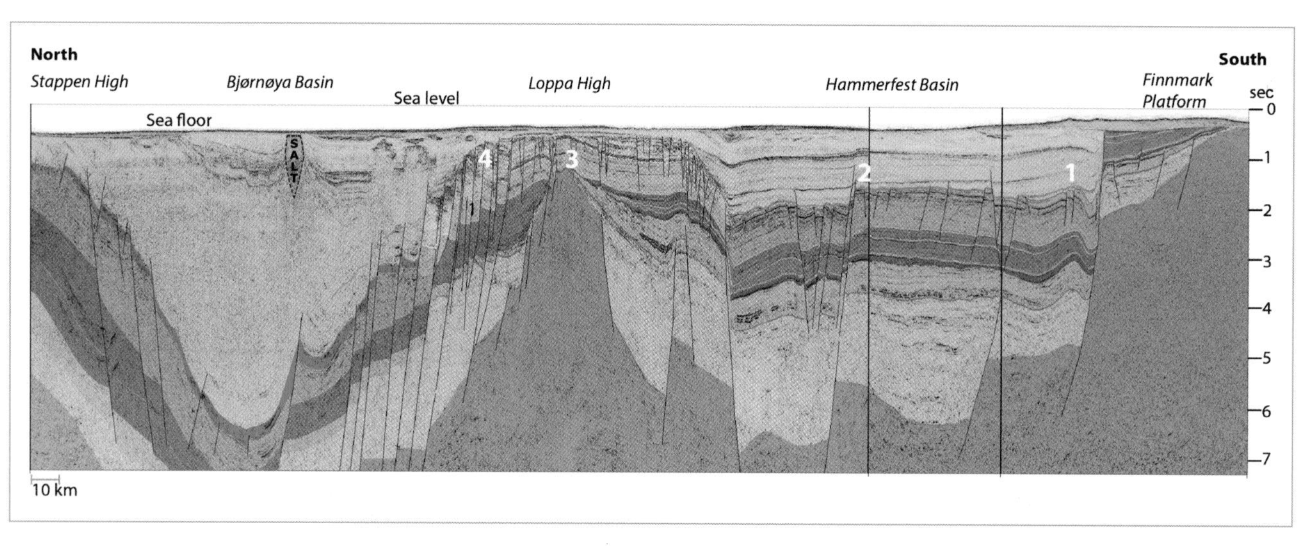

Cross section of the upper layers of the earth's crust in the southwestern Barents Sea. These exemplify sedimentary basins: the Bjørnøya and Hammerfest basin and adjoining heights. The numbers on the right give the seismic two wave travel time, in seconds. In this section, 7 seconds corresponds a depth of about 14 km under the seabed. Green denotes Caledonian or crystalline basement rocks like those in mainland Norway. Orange and beige indicate shales, limestones and sandstones from Late Paleozoic. Mesozoic rocks are in purple, blue and light green, and Tertiary rocks are yellow. The orange part at the top is sand, gravel and clay, glacial deposits from the most recent ice ages. White numbers mark traps/structures containing major discoveries of oil or gas. As the figure shows, oil and gas accumulate on highs or along the flanks of the basins. The number 1) shows a structure similar to the Goliat field (the traps are about 700 m to 1.5 km under the seabed) which holds discoveries in Triassic strata; 2) shows the position of the Snøhvit gas field (traps about 2 km under the seabed) where the reservoir consists of Jurassic sandstones; 3) shows the position of the Gohta and Alta discoveries, which are in Permian carbonate rocks; 4) shows the geological position of the Johan Castberg field, where the oil accumulations are in Jurassic and Triassic strata about 1–1.5 km under the seabed.

The Norwegian Petroleum Directorate's classification of resources and reserves

The classifications used by the Norwegian Petroleum Directorate (NPD) to calculate resources and reserves closely match those recommended by the Society of Petroleum Engineers and the United Nations. Simply put, NPD subdivides oil and gas into reserves, contingent resources and undiscovered resources:

1. *Reserves: petroleum remaining in fields approved for production.* Those closest to us are the Snøhvit gas field, and the oil fields Goliat, and Prirazlomnoye across the Russian border.

2. *Contingent resources: known quantities of petroleum for which additional clarification and decisions are needed before they can be approved for production.* Examples include the new discoveries Johan Castberg, Gohta, Alta, Wisting, and Hanssen in the southwest Barents Sea. The gigantic gas field Stockmann on the Russian side is assigned to this category in the NPD system, because no decision has yet been made to go ahead with exploitation, whereas the Russian system classifies Stockmann as a reserve.

3. *Undiscovered resources: petroleum resources that have not yet been found, but that are assumed possible to discover, and that might be exploited in the future.* In the Norwegian Arctic, examples include Lofoten–Vesterålen and areas north of Bjørnøya.

The best known and most frequently cited overview of arctic petroleum resources north of the Arctic Circle is the one from USGS. It uses the term *undiscovered conventional oil and natural gas resources*, and the estimates fall under category 3 of the NPD classification system – undiscovered resources: petroleum resources that have not yet been found, but that are assumed possible to discover, and that might be exploited in the future.

Petroleum resources in the Arctic

For several of the sedimentary basins in the Arctic, important data have not yet been collected – information that is required to predict the presence of an effective hydrocarbon system. In some cases, the geological data are so limited that they are not used in assessments of a basin's potential to produce conventional oil or gas. In addition, in several areas, the geological and geophysical data that have been collected are neither published nor publically available. When making an estimate of conventional petroleum resources in the Arctic, it is important to remember the words of Piet Hein: *Knowing what thou knowest not, is in a sense omniscience.* Classification and naming of basins and provinces in the Arctic is controversial.

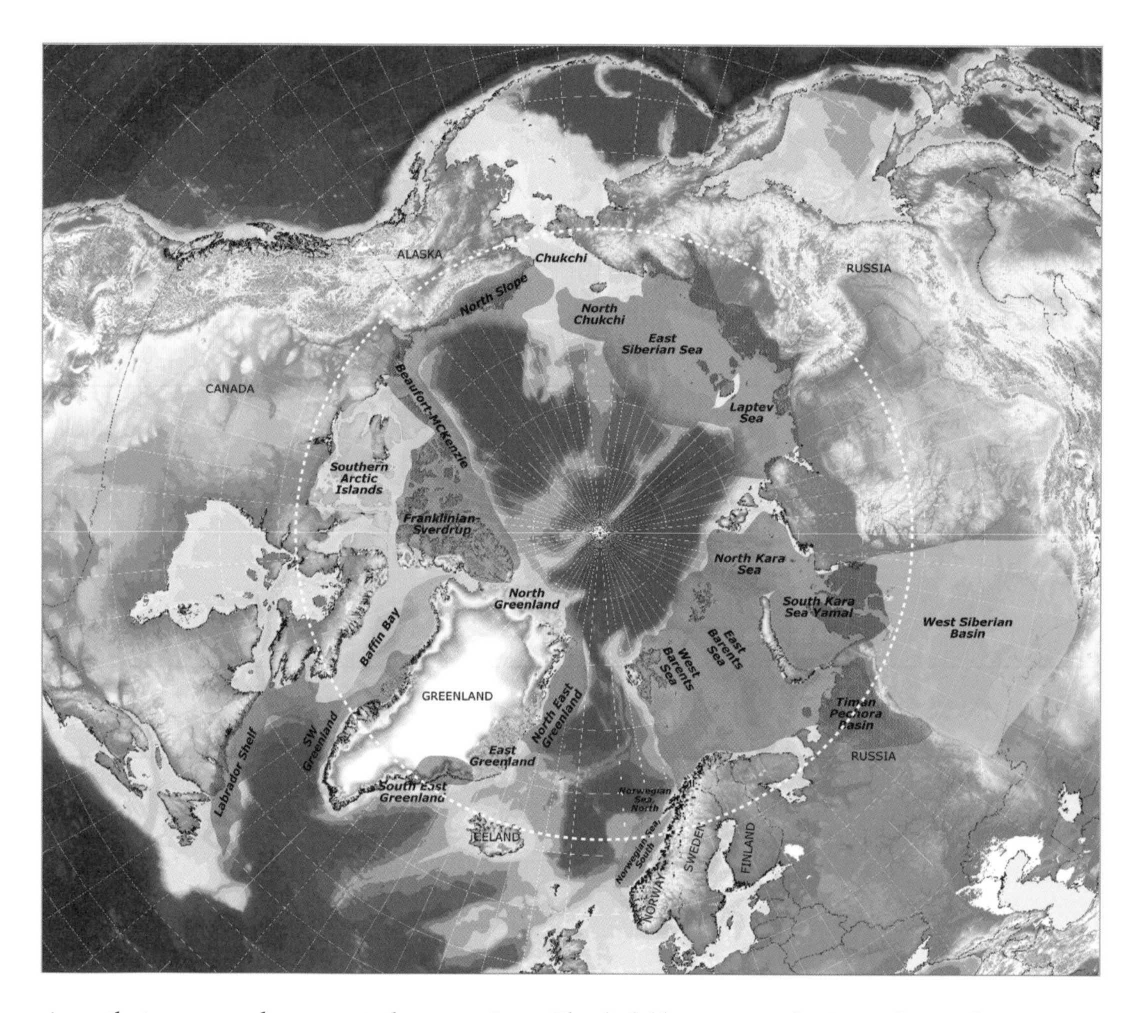

Areas that are or can become petroleum provinces. The dark blue areas are basins without sedimentary rocks. This means that the deepest parts of the Norwegian Sea, the Greenland Sea and the Arctic Ocean consist of oceanic crust and are not of interest for prospecting for conventional hydrocarbons. In the other areas north of the Arctic Circle, USGS has estimated about 14.3 billion Sm^3 (90 billion barrels) of extractable oil resources, 7 billion Sm^3 (44 billion barrels) of extractable liquefied natural gas resources, and about 47.7 trillion Sm^3 (1668 trillion cubic feet) of extractable gas resources. Figure adapted from Olaussen & Steel (2011) Getting started in the Arctic. A compilation of selected papers from the sedimentary basins in the arctic region. American Petroleum Geologists AAPG Getting Started Series (CD).

The figure on page 118 shows an informal attempt at classification, along with a simplified classification of geological/geographic provinces with sedimentary basins in the Arctic. In general, they can be described as current or possible future petroleum provinces.

USGS estimates of exploitable resources in arctic basins

In its impressive published overview, the Geological Survey, USGS, states that we can expect 14.3 billion Sm³ (90 billion barrels) of exploitable oil, 7 billion Sm³ (44 billion barrels) of liquefied natural gas, and 47.7 trillion Sm³ (1668 trillion cubic feet) of gas. In total, this is 55 billion Sm³ oil equivalents. Of this, 84% is offshore, most of it at depths shallower than 500 meters. In the same publication, USGS writes that this is the "best estimate", with an extremely large range of uncertainty, both up and down. USGS was the first to clarify the uncertainty of these estimates. Unfortunately, the uncertainties are not mentioned by the media.

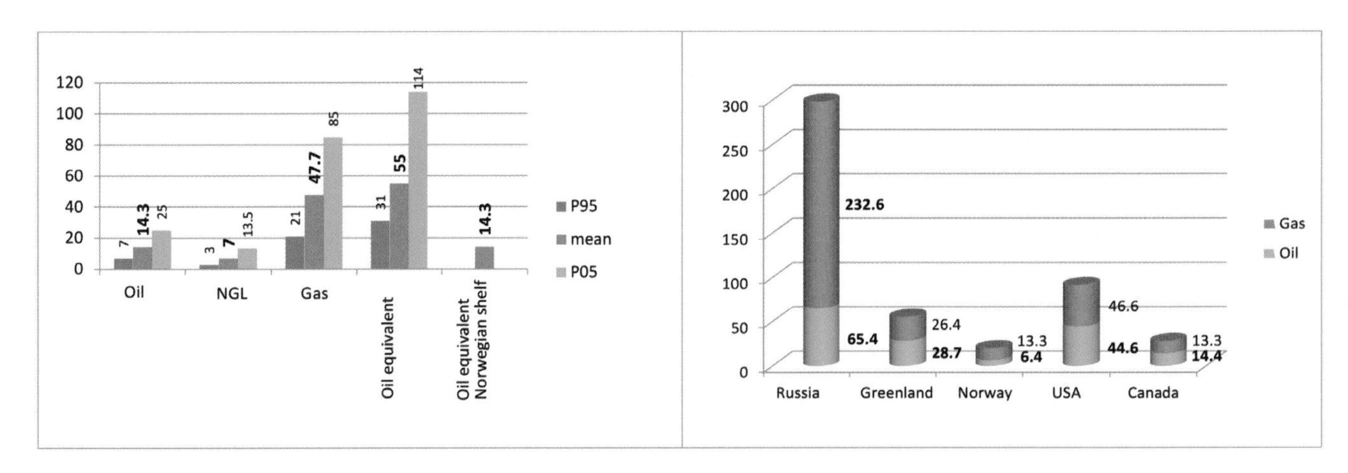

Sample inventory of resources in the Arctic. The figure on the left shows USGS estimates (with margin of uncertainty) of oil and gas resources north of the Arctic Circle. "P05" indicates a 5% probability of finding the high estimate, and "P95" the likelihood that the low estimate will be found. These numbers are compared with expectations for the Norwegian arctic shelf. NGL – natural gas liquids. The figure on the right is from an overview by V. Vladimirov (2013) of offshore oil and gas in the Arctic. Oil and oil equivalents are in billion standard cubic meters (10^9) and gas is in trillion standard cubic meters (10^{12}).

Alaska, Canada, Greenland, Siberia, and the Russian Barents Sea sector are most promising
USGS has also estimated that 70% of the exploitable oil north of the Arctic Circle is in the Alaskan Arctic, north of Sverdrup and Beaufort (the Amerasian Basin), northeastern Greenland and western Greenland/eastern Canada. Most of the natural gas is assumed to be in arctic areas of Alaska, in the western part of the Siberian basin, and in the eastern Barents Sea. Russian authorities claim that Barents Sea East, along with the Kara Sea, has the greatest petroleum potential.

So how much can we trust these numbers? Even though the estimates are highly uncertain, a basic understanding of geological processes and the basins' proximity to more mature hydrocarbon provinces allow us to make qualified guesses about the resources they hold. Nevertheless, we should once again remember Piet Hein's words, and add that the only thing we know for certain, when we hear a specific value for oil or gas resources in an arctic region or in any frontier basin, is that the number is probably riddled with error. Even in a mature area like the North Sea, where prospecting has been going on since 1965, the margin of uncertainty is wide. If the goal is to reduce uncertainty linked to estimates of resources in the Arctic, we need more geological and geophysical data and knowledge – and more boreholes.

The Norwegian Arctic

The Petroleum Directorate gives the following numbers (discovered and undiscovered recoverable resources) for the Norwegian part of the Barents Sea: 589 million Sm^3 oil, 97 million Sm^3 gas condensate and natural gas liquids, 1046 billion Sm^3 gas, or 1739 million Sm^3 oil equivalents. Are these figures overly pessimistic?

Has the western margin of the Barents Sea been underestimated?
In 2008, USGS did not consider the Norwegian parts of the Barents Sea particularly promising, as can be seen in the figure on page 121. If we look more closely at the western margin of the Barents Sea (which USGS calls the Norwegian Margin), USGS considered this mainly as a gas province. The western margin, as defined by USGS, is indicated in the

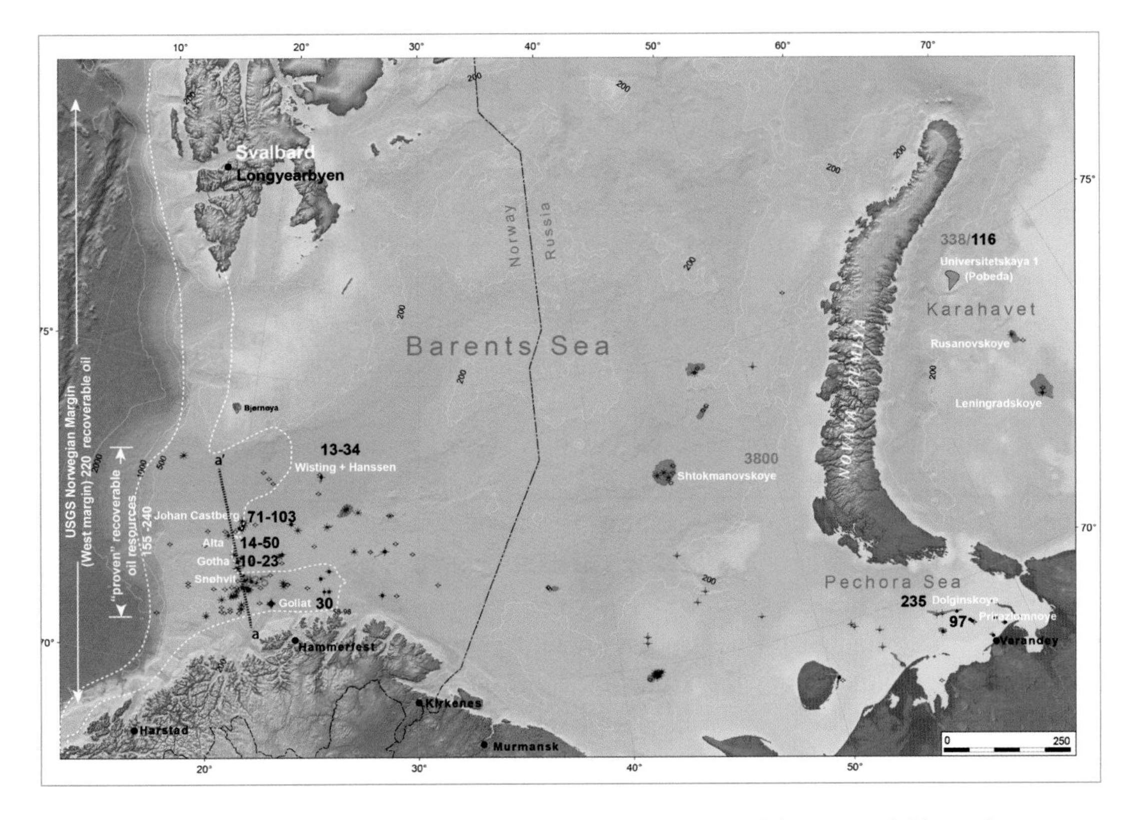

Map of the seabed in Barents Sea and adjacent land areas. Important oil discoveries/fields are shown in red and gas discoveries/fields in black. The numbers show total recoverable resources and reserves in fields prior to production. The figures are from NPD and indicate the range of the estimates (lowest to highest) for individual discoveries and fields. All figures for oil are given in million (10^6) Sm^3, and those for gas in billion (10^9) Sm^3. Source: NPD and Rosneft/Gazprom. Modified map from the Norwegian Polar Institute.

same figure. Discoveries of oil in recent years show that the prognosis for this area was too pessimistic.

Prospecting on the western margin of the Barents Sea in the past five years has yielded the interesting oil discoveries Johan Castberg, Gotha, Alta, Wisting, and Hanssen. These discoveries, along with improved geophysical technology and greater insight into the area's geology, give reason for much more optimism concerning the western margin.

When USGS was estimating resources in this area, only the gas field Snøhvit, with a thin oil zone, the upper reservoir unit of the Goliat field, several non-commercial gas discoveries, and one non-commercial oil discovery were known. USGS set an estimate of just 220 million Sm^3 exploitable oil resources, while the gas resources were estimated at about 1100 billion Sm^3.

Time has shown that the prognosis of a gas province with negligible potential for oil was incorrect. If we include the categories *reserves* and *contingent resources*, over 150 million Sm^3 of recoverable oil resources have already been found. In other words, after just a brief span of renewed prospecting, the discoveries have *already* nearly lived up to mean expectations.

There is therefore reason to believe that the USGS figures for oil in the western margin are too pessimistic; it is reasonable to expect double the USGS value. This is supported by the fact that considerable oil accumulations have been proven in new play models that can be expected to extend to several other parts of the western margin. In addition, new geological knowledge – particularly improved understanding of fluid dynamics in the rock bed and better geophysical methods – will probably raise the estimates.

Uplift of the shelves provides "new oil"

An important geological factor in the Barents Sea – in fact in many of the arctic basins – is that the basins have been uplifted considerably over the past 2.5 million years. In some places, the total uplift was over two kilometers, and gave subsequent erosion of shelf sediments.

Svalbard, which is the exposed part of the bedrock of the Barents Sea, has risen most, and the upward movement continues. The Norwegian Mapping Authority has measured an annual uplift of 8.2 mm in Ny-Ålesund. The rate of this rise has not been constant. During ice ages, the crust has been pushed down, and when the ice retreated, uplift resumed and erosion took place. Fridtjof Nansen wrote about this as early as the beginning of the 1900s. He charted, on the bottom of the Barents Sea, rivers and flood plains, signs that erosion had shaped sediments laid down earlier. The Barents Sea is submerged tundra with mighty riverbeds and lakes, not unlike parts of present-day Siberia.

The uplift that follows the retreat of glaciers and the subsequent erosion lead to phase changes in oil that has accumulated in deep traps: when the pressure on oil is reduced, it fractions into oil and gas. This takes more space; some of the oil is therefore pressed out and seeps upward to shallower traps or to the seabed. In addition, movements in the earth's crust can alter flow patterns in the bedrock; traps tip to one side and leak oil, which seeps upward. Another possibility is that new oil is formed as the source rocks mature and expel oil during basin subsidence in the west. This means that the flanks of the basins are interesting for potential accumulation of oil. The Goliat field, and the discoveries Johan Castberg, Gohta, Wisting, Alta, and Hanssen are all in this type of geological position (see figure on page 116). They are placed so they lie within current efficient petroleum systems.

A new area in the Norwegian Barents Sea – most likely a gas province?

So what about the new Norwegian area at the southern end of the Barents Sea, along the previously disputed border between Norway and Russia? In the southeastern Barents Sea, NPD expects about 300 million Sm^3 oil equivalents, mainly in the form of gas.

Almost without exception, the media describe this as vast oil and gas reserves in the southeast. But no, these are disappointing figures for a region nearly as extensive as the northern North Sea. The fields Statfjord and Ekofisk each have reserves twice as large. The new Norwegian area in the southeast Barents Sea cannot be compared with the northern North Sea as it lacks the world class Upper Jurassic source rock and the highly favorable trap configurations contained in the prolific North Sea oil province. But we can choose to be a bit more optimistic concerning the likelihood of finding more oil, since this area is on the flank of a plateau – the same kind as in the western margin of the Barents Sea with a deeper basin below. Insight and experience from the west allow us to take a more positive view. In addition, we are now aware of the existence of potential source rocks at several different levels: they can be seen both in Svalbard and in boreholes in the Barents Sea. Again, as in the western margins, oil that has accumulated earlier may have seeped up to shallower accumulation traps.

Preconceived notions in prospecting

Preconceived notions are nothing new to petroleum exploration along the Norwegian shelf. In the early 1980s, the eastern part of Norwegian Sea was considered a gas province. This attitude changed quickly when the gigantic Heidrun field was discovered. This part of the

Norwegian Sea is now a rich oil and gas province, containing not only Heidrun, but also the fields Åsgard, Draugen and Norne, classified as giants.

The resource potential in a new or unexplored area is often compared with that of geologically analogous productive basins. Play models that have proven useful in the basin already in production are often transferred to the new, unexplored basin. Until oil was discovered in the Goliat field, the successful models from the Norwegian Sea and, particularly, the northernmost North Sea were used as analogues for evaluation of prospects in the southwestern Barents Sea. When the oil discoveries there were disappointing, the Norwegian sector of the Barents Sea came to be considered a gas province, and the uplift was seen as an obstacle for oil accumulation. Geologists know better now. Does this mean that the prediction of a possible gas province in the southeast will turn out to be incorrect?

Intensified activity in the Arctic? Yin and yang

Yin

Suppose we take at face value the USGS figures stating that 14.3 billion Sm^3 of exploitable oil remains to be discovered in the Arctic: this is presented as a huge volume in the media and among politicians. But is it really?

Looking exclusively at oil, this corresponds to the world's total consumption over a three-year period. It is also less than what each of the nations Iraq, Kuwait, the United Arab Emirates, and Iran claim to have as exploitable reserves.

To reiterate, unconventional petroleum is not included in these calculations. For comparison, little Norway has estimated that the total recoverable potential oil sources in all Norwegian offshore areas comprise 7.2 billion Sm^3.

Logistical, environmental and economic challenges

In their estimates of undiscovered oil and gas resources, USGS has not taken economic factors into consideration – factors such as year-round sea ice and water depth. Of the 14.3 billion Sm^3 of exploitable conventional oil resources, much will probably remain "untouched". The reason for this is that operating conditions will be tricky, involving ice, darkness, and prohibitive transportation costs. In addition, according to USGS, there may be a dearth of pertinent technology and qualified labor. But the most important obstacle of all will probably be the obligation to adhere to strict, costly environmental restrictions.

An example of this last point is that petroleum operations are banned in large segments of arctic Alaska. This is an area where the geology is well known and the estimated resources vastly exceed those predicted in Lofoten/Vesterålen. All five arctic countries with offshore petroleum operations have agreed on restrictive environmental policies. Additional areas may therefore be protected in the future.

Considering the operational challenges, the addition of 14.3 billion Sm^3 will not have any major effect on the world's energy situation.

Relatively modest "Big Oil" activity in the Arctic

Even today the challenges of the Arctic are reflected in the major international and national energy conglomerates' relative lack of interest. Apart from in Barents Sea South (Norwegian sector), Alaska, the Hope/Chukchi area (west of Alaska) and parts of the western Russian Arctic, there is clearly very little interest at present from the big international energy companies, also known as "the Majors", "Supermajors" or "Big Oil".

The focus is currently on the United States (shale oil and gas), the Gulf of Mexico, Canada (oil sand), Central and South America, Africa, the Caspian Sea, and Oceania. In other words, many of the oil resources that have been identified in the Arctic are at present too costly and too risky to warrant a geographical shift in prospecting efforts.

Yang

Which regions are considered to be of interest for prospecting can shift rapidly if a new major field is found, or if global demands for energy security grow even larger. Despite the picture sometimes presented in the media – of supposedly bitter border disputes, touchy political situations and military buildup – the Arctic is a thoroughly regulated and politically stable region, where dialogue has been on the agenda, rather than conflict. This is in stark contrast to many of the countries with large reserves of conventional and "cheap" oil. Thus oil prospecting and production in the Arctic is politically predictable for the oil companies.

I believe it is correct to say that all five countries bordering on the Arctic Ocean have an intrinsic and genuine interest in sustainable development of mineral, petroleum and food resources in the Arctic. I am naïve enough to believe that this is also in the interest of the energy conglomerates. So despite operational challenges, high costs and strict environmental regulations, we will probably see increased activity, at least in ice-free areas – and in a few ice-covered areas.

This belief is supported by the following factors, in addition to the ones mentioned above.

- The industry knows that huge fields have been found in the Arctic, such as Prudhoe Bay on North Slope in Alaska, which originally held oil reserves exceeding 2 billion Sm^3. For comparison, Norway's two largest fields, Ekofisk and Statfjord, hold 0.6 billion Sm^3 each. Stockmann, in Barents Sea East, has over 4500 billion Sm^3 gas (130 trillion cubic feet), more than three times as much as the giant field Troll. The gas fields at Jamal, partly on land, partly offshore along the southern coast of the Kara Sea, have over 12,000 billion Sm^3 gas (350 trillion cubic feet). So some areas are clearly interesting.
- New exciting oil discoveries in the southwestern corner of the Norwegian sector of the Barents Sea, i.e. the Johan Castberg, Gotha, Alta, Wisting, and Hanssen discoveries.
- A new exciting area between Norway and Russia is being opened up on either side of the border.
- The world's northernmost oilfield, the Eni-operated Goliat field, will soon begin production in the southwestern Barents Sea. Oil production has started from the gigantic Gazprom-operated oilfield Prirazlomnoye in the Pechora Sea, in the southeastern margin of the Barents Sea. Both examples show the economic and logistical feasibility of extracting resources far in the north.
- In 2013, licenses were issued for prospecting offshore of northeastern Greenland. According to USGS, this is one of the most promising areas in the Arctic.
- Offshore drilling, new licensed areas, and major discoveries in the Canadian part of Baffin Bay, near western Greenland.
- Renewed interest in Alaska and particularly the Chukchi Sea.
- A new exciting giant discovery in the Kara Sea.

- Areas available for production of conventional “cheap oil” are shrinking because of nationalization, political unrest, and stiffer competition for new prospecting areas, not least from China. Basins expected to contain “cheap oil” are now assumed to be few and will become even fewer because of national interests.
- Major industrial nations (e.g. the European Union, Japan, China, and India) are placing greater emphasis on energy security.
- New and better technology will make exploration and production cheaper and safer.
- Based on the world’s sharply growing demand for energy, there is good reason to predict increased activity in the most easily accessible arctic regions in the future. In the first instance, this means the ice-free areas of the Arctic.

Svalbard and petroleum resources

Exceedingly promising source and reservoir rocks are known to exist in Svalbard. The eastern and southern parts of Spitsbergen and the islands to the east and south have outcroppings of rock similar to those that contain oil and gas in the Barents Sea. Several limestone and sandstone strata are stained with oil, and natural seepage of oil and gas from the bedrock has been observed. Non-commercial oil and gas discoveries have been confirmed in several boreholes on Spitsbergen. But locating conventional oil and gas has hitherto entailed so great an economic and environmental risk that the oil companies have not been interested. Judging from present knowledge, economically viable discoveries of conventional oil are unlikely in Svalbard.

Unconventional oil and gas prospects in Svalbard?

Conversely, discoveries of unconventional oil and gas appear highly promising. Both shale gas and shale oil are probably present, and the coal-mining company Store Norske has demonstrated the possibility of extracting considerable volumes of oil transformed from coal. But again, this has so far been of little economic interest owing to transport costs and – not least – desirable environmental restrictions.

Svalbard's geology increases understanding of hydrocarbon resources in the Arctic

Svalbard holds keys that can help unlock the geological secrets not only of the Barents Sea, but also adjacent arctic basins in Russia, Greenland, and northern Canada. Better knowledge about Svalbard's geology will improve estimates of oil and gas resources in these areas, and simultaneously make oil prospecting in the Arctic a less risky endeavor.

Ethical considerations related to new economic activities in the Arctic

by Gunnar Sander, Senior advisor, Norwegian Polar Institute / PhD Candidate, UiT – the Arctic University of Norway

Climate and nature in the Arctic will undergo major changes in the years to come, even if an effective global climate agreement delivers us from the most devastating negative effects. Expanding commercial enterprise will put the arctic environment under additional pressure, but may also provide income and positive ripple effects. While most of the chapters in this book deal with ethics and climate, this one will focus on some of the ethical dilemmas posed by expanding industry.

There is no simple, neutral answer to the question of right and wrong in this context. The effects are unevenly distributed: some people benefit and others suffer. While a mining operation makes a profit for the owners, waste dumped in the fjord may threaten the livelihood of those who harvest its fish and shellfish. The same change can affect different groups in completely different ways: less ice means easier passage for ships, but those who hunt seals on the ice see the basis for their time-honored hunting culture literally melt away. Seal hunting is in itself a good example of how cultural values about a phenomenon can collide: is hunting a natural way to harvest nature's bounty or unethical cruelty to animals?

The point is that our interests and values shape our perceptions of what is positive and worthy of support, and what is a problem and requires action and solutions. From diverse attitudes it is not far to politics, where society discusses what kinds of commercial activities should be stimulated or stopped, and what steps must be taken to prevent unwanted consequences. These are sweeping debates full of technical details. But the opinions voiced reflect certain basic values that we will present here, and that deserve to be discussed explicitly.

What is the value of the arctic environment?

Humans are making their mark on an ever-increasing proportion of the globe. We impoverish ecosystems so quickly and extensively that it is beginning to lash back on human economy and welfare. Even the most remote parts of the earth feel the effects of our activities in the form of emissions that change the climate, toxic substances that follow air and ocean currents, and perturbations in populations of migrating animals.

The Arctic is one of the areas where humans leave very few direct imprints and where natural processes still hold sway. The indigenous peoples who have lived here for millennia have been few; their traditional way of using nature has involved simple technology, and few places have been affected by modern industrial society.

What is the value of this relatively untouched arctic environment? One way to approach this question is utilitarian, focusing on what types of goods and services the ecosystems offer to humans. The snow and ice that cover the arctic landscape have global importance owing to their ability to reflect heat from the sun and keep the climate favorable. This is a good example of a regulatory ecosystem service. Ecosystems also provide us with products: food, animal skins, useful molecules, firewood, and hydrocarbons; wind and minerals too, if we include the inorganic parts of ecosystems. A third type of service is cultural and can involve experiences in the wild, spiritual enrichment, research, and increased knowledge. People who have never been in the Arctic value knowing of the existence of a unique arctic environment. Independent of these anthropocentric utility considerations, many ascribe to nature an intrinsic value of its own that must be respected.

It is quite obvious that viewpoints diverge concerning how valuable the arctic environment is, and what types of assets are seen as most important. Those whose values support restrictive attitudes toward economic exploitation of the Arctic often emphasize the uniqueness and irreplaceability of the arctic environment, and that pristine nature is a rare commodity that will increase in value as it continues to vanish in other places. The same views are supported by those who accord greater importance to cultural and regulatory ecosystem services than to the Arctic's ability to offer products. Arguments are also often voiced to the effect that it is ethically correct to ensure that opportunities are shared between generations: our generation should not help itself to more of the resources than our fair share, and should not degrade nature in ways that limit the options of our descendants. These values are incorporated into the concept of sustainability as presented by the Brundtland Commission. How they should be applied to the extraction of non-renewable resources in the Arctic remains debatable.

Values that support increased exploitation of arctic resources often view nature's ability to supply products as its most important service; it is directly useful in creating employment, income, and human welfare. Immediate benefit today can be accorded greater importance than long-term benefit for our descendants. Sometimes one can come across the argument that it is unimaginative to judge future possibilities on the basis of current realities, because humans have always found new ways to adapt. Another type of discourse makes the value of arctic nature relative. Yes, it is unique, but other ecosystems in other places also have unique adaptations to their local surroundings. Regulations, technology and other conditions may be more advantageous in the highly-developed countries of the Arctic. It is best to use nature in the places where exploitation will do least damage, so we can harvest in the Arctic when it is most advantageous.

What demands can we place on commercial enterprise?

The dominant view on exploitation of arctic resources is that it is unrealistic or undesirable to block all increases in activity. Policy declarations from the countries that have a formal policy on the Arctic and from various organizations share a common feature: a stated wish both to use the region and to protect the vulnerable environment. There is a search for balance, a golden medium that can be defined as sustainable. But sustainability is a broad concept. It must cover ecological, economic and social conditions, and tackle tricky themes like resource distribution over generations and between rich and poor in our own time. Diverse value judgments are possible; sustainability means different things to different people, and use of the word may end in a battlefield of disparate interpretations.

Other principles are more clearly defined and are thus easier to use as guidelines for practical action. One of these is that the total pressure on ecosystems should determine what activities are permissible. This is not always easily accepted by developers who feel that their own activities have very little impact. Climate change and ocean acidification may change arctic ecosystems dramatically, and we know that certain species at the top of the food chain are harmed by pollutants transported from far away. If the ultimate goal is to preserve the basic characteristics of arctic ecosystems, the scope for additional activity in the region may be limited.

Taking precautions is another important principle. How this is expressed varies, but the core concept is that if there is any doubt concerning the risk, nature should have priority: even if we are unsure about whether the environment will actually be threatened, this should not prevent us from taking steps to prevent harm. Uncertainty is more the rule than the exception; scientifically speaking, we can never be certain that our hypotheses are true, but for practical purposes, we must accept some knowledge as true. But how stringently should we evaluate the probablility of harm before we decide that countermeasures must be taken? This is a matter of how we relate to risk. Some commercial interests demand that a potentially harmful effect must essentially be proven beforehand, whereas environmental organizations accept poorly founded hypotheses as good enough reasons to ban an activity or demand countermeasures. Costs and sharing of burdens also factor in. One example is ships sailing the Northeast Passage with alien species that could potentially wreak havoc on existing ecosystems. There are still no regulations in force demanding that ballast water be cleaned or that hulls should be scraped. Should shipowners operating in the Arctic be forced to pay extra costs to cut risks – even before global agreements are in place and all shipowners must pay the same costs? If shipowners are forced to shoulder responsibility, is that an appropriate and just application of the "polluters-pay" principle? Would it be fair to say that those who benefit from intact ecosystems should also contribute?

Other relevant principles in this context include doing environmental impact assessments, using the best available technology, and that the polluter should pay. Implementation of these principles can lead to specific environmental protection measures, such as prohibiting certain types of activities, separating activities and vulnerable environments in time and space, or strictly regulating emissions, risk levels or technologies. There will always be compromises concerning what regulations are required, and how rules should be interpreted. These discussions often appear technical and incomprehensible. But behind a choice between different solutions, there is often a value judgment about how a principle should be implemented in practice, and which other considerations should be taken. These questions need to be raised in open forums and in plain language so public debate is possible.

Who should the Arctic be developed for?

Historically, the Arctic has been an outpost, its resources exploited by powers far away, as if it were a colony. It is easy to interpret much of the current global and national interest in a similar light: the Arctic is of interest primarily because it can confer resources and wealth on those willing to invest there and on the governments that take their share of the profits. But the region's inhabitants are also raising their voices to say they want to be part of the development. Canada, in particular, has demanded that development in the Arctic benefit the inhabitants. This must be seen against the backdrop of poor living conditions for the Inuit, but is also true of other indigenous peoples and regions in the Arctic.

The degree to which resource exploitation benefits local communities and economies in the Arctic depends on how the activity is organized. Mineral mining and petroleum operations are often run from corporate headquarters far away. The workforce may commute to a "company town" or oil platform and go home at the end of each shift. Raw materials may be shipped without being processed locally. Fast exploitation of a resource can lead to "boom and bust": rapid expansion and a flurry of activity followed by shut-down once the resource has been depleted. But these projects can also contribute to long-term regional evolution, provided the developer enters into dialogue with the community, strives to raise local employment rates and competence, and to improve infrastructure, health services and schools. Ideally, this will lay the foundation for other activities and more long-term development. Within fisheries, the issues of resource ownership and where the catch should be landed have major impact on who pockets the income and who benefits from spin-off effects: a seagoing fleet where the catch is frozen for export, or local fishermen whose catch sustains local processing enterprises? These examples show how different ways of organizing economic activities can have contrasting socioeconomic consequences on both local and regional scale.

Demands for development strategies that foster spin-off effects in the local economy can raise costs for the developer; on the other hand, such strategies can be socioeconomically beneficial. Discussions often weigh advantages and disadvantages against each other. For example, if northern Norway bears the risk and inconvenience of oil production, the region ought to benefit in terms of more jobs and other spin-off effects. The discussion holds strong elements of what is fair and socially acceptable. This is clearly a topic with ethical dimensions.

Who gets to decide?

Many have interests in arctic development. In the arctic countries we find ordinary citizens, commercial enterprises and a variety of organizations, in addition to elected bodies that administrate at local, regional, or national level. All the arctic countries acknowledge the rights of indigenous peoples, but the political approaches range from autonomous regional self-government to powerless organizations under controlled democracy. The eight arctic states are central players, but many other countries also have interests here. A range of international organizations are active, including inter-governmental bodies and non-governmental organizations. Other interest groups are more difficult to identify and find representatives for, such as "future generations" and "the international community".

Which rights and obligations do these players have, and which of them should be involved in decision-making?

Legal statutes provide some guidance. The arctic states have sovereign power within their territories, which extend 12 nautical miles from their baselines (drawn along their outer shoreline). Beyond that, the main rule is that the coastal states control natural resources out to 200 nm, or farther if there is an extended continental shelf, but that other states have a right to engage in activities that are not related to exploitation of resources, such as shipping. Beyond the coastal states' maritime zones lie the High Seas and the deep ocean seabed (see map). These are international spaces where all countries have equal rights. The deep ocean seabed has even been declared "common heritage of mankind" and is managed by a UN Agency in Jamaica.

The arctic states control most of the region and naturally play a major part in governing the Arctic, including the ocean. But they cannot monopolize management and ignore the rights assigned to other states under the law of the sea. The challenge is to find ways to cooperate that give both the arctic states and others balanced influence over specific issues

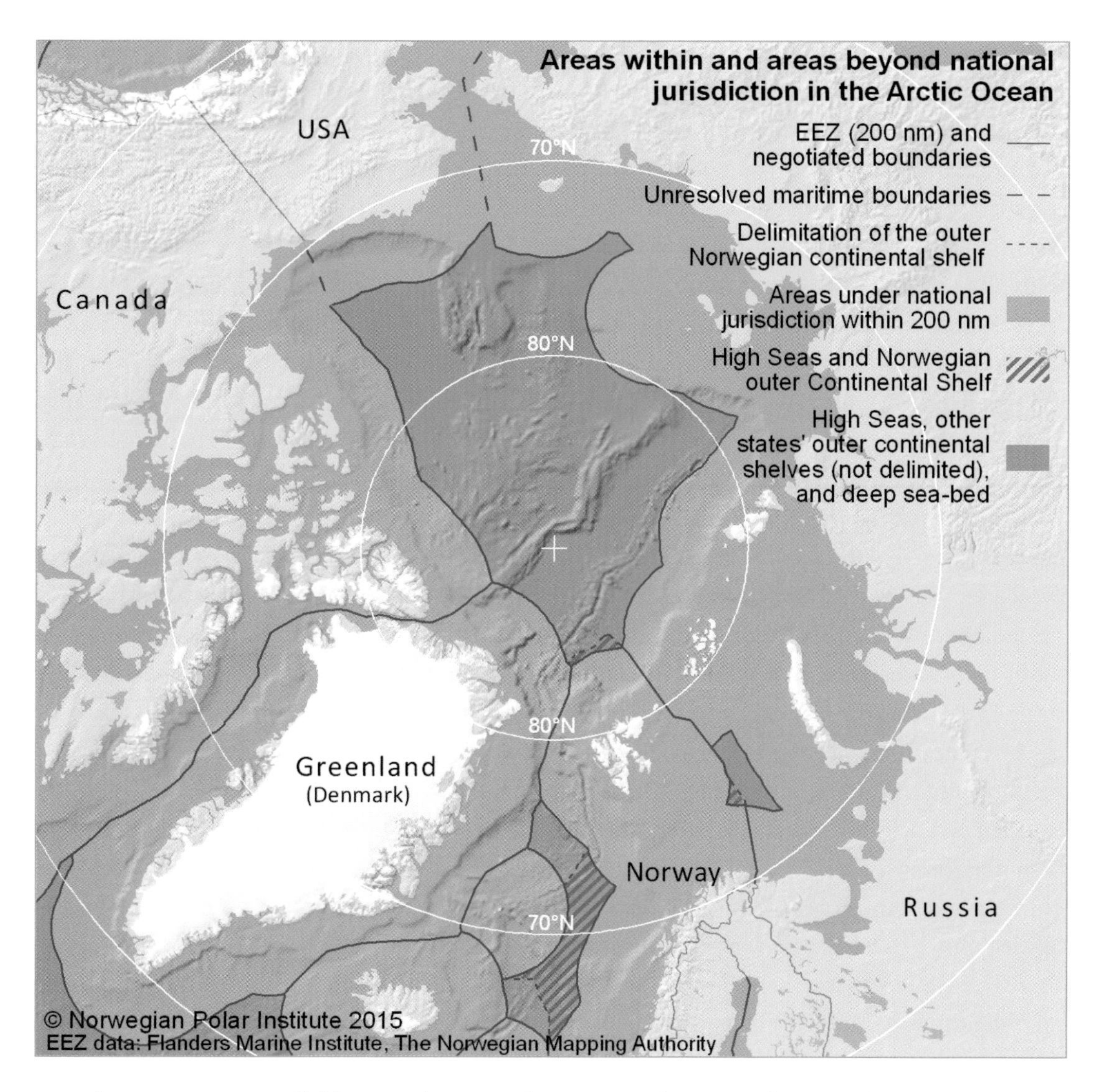

Map showing areas controlled by coastal states and international areas: High Seas and the deep ocean seabed beyond the continental shelves. So far, Norway is the only state for which the United Nations' Commission on the Limits of the Continental Shelf has recommended a shelf boundary outside the 200 nm Exclusive Economic Zone (EEZ). All the coastal states must have their extended continental shelf delimitations evaluated, and negotiate delimitations between them, before this issue is settled. An extended continental shelf does not change the overlying waters' status as High Seas. (Illustration: Norwegian Polar Institute)

and areas. Several solutions are possible. One example is the ongoing negotiations within the UN's International Maritime Organization to establish a binding Polar Code for shipping. But the law of the sea also gives arctic coastal states the right to regulate shipping within their 200 nm zones in ice-covered waters, trumping international regulations if they see fit. The question is who can place demands on ships: coastal states, which traditionally want to protect themselves against potential harm from shipping, or a global forum, which may look more to the flag states' desire for uniform regulations and the low standards of countries that offer "flags of convenience"? Management of potential resources for fisheries in the High Seas of the Arctic Ocean is another example. All states have the same right to these resources, but the management must take into account policies in the coastal states. A fishery regime may include only the High Seas, or the entire Arctic Ocean; this will impinge on who should have most influence. The arctic coastal states have initiated a process to find a solution for the High Seas, so far without involving other countries (see the chapter by Alf Håkon Hoel).

But perhaps the environment and the resources in the Arctic are important enough to be of global concern? The international spaces definitely are, regardless of what value they are assigned. What about the areas under national jurisdiction? The arctic states relinquish a bit of their sovereignty every time they sign an international agreement and pledge to abide by it. They then become accountable to other parties to the agreement, for example concerning how they manage migratory species or assess the risks of new activities that might affect neighboring countries. This means the international community will have greater influence over the Arctic if more arctic states ratify existing conventions or sign new agreements with new obligations. Customary law has the same implications: climate, biodiversity, and the environment can no longer be seen as purely domestic affairs. However, a state's responsibilities under customary law are harder to define than its responsibilities under treaty law.

The arctic states can also make commitments to the international community by taking on obligations voluntarily. Norway's involvement in efforts to persuade Brazil and Indonesia to save their rain forests is an example. Most Norwegians probably see this as legitimate involvement in a globally important issue, and believe our engagement strengthens Norway's image as a moral superpower – a self-image we take pride in. But are we equally pleased about foreign involvement in what we consider our own domestic affairs – or would we find it more important to assert sovereignty and self-rule?

Voluntary involvement brings us to other stakeholders. Commercial enterprise is central in giving impetus to increased activity in the Arctic. Heavy responsibility lies on those who exert pressure to open new areas, for example for oil exploration, and companies are influential. They can be cautious in choosing what they get involved in, and can also choose to adhere to higher environmental and social standards than those set by the authorities. Ethics and corporate social responsibility in business is currently in vogue; many companies formulate ethical guidelines and wish to achieve a positive public reputation. Their practical actions must then tolerate public scrutiny and debate concerning whether they measure up to their own standards and those of others.

Within a country, laws define the balance of power between state, regional and local authorities, and the rights of citizens. Just as in the international system, national management evolves constantly as new issues arise. Who makes and who influences those decisions is determined by considerations involving rights, fairness, proximity and efficiency – quantities that are difficult to measure objectively.

Who makes decisions is one question. It is equally important to ask what foundation decisions about commercial development in the Arctic should be based on. It should be a requirement that several alternative development strategies are presented, and that their possible impact is reasonably well assessed. Impact assessment is a key instrument that requires extensive public involvement. Important issues should be brought to the public for an informed discussion of advantages and disadvantages and which values should count most. This is democracy.

Epilogue

We must expect that pristine parts of the Arctic will increasingly be put to use. When this happens we must ensure that we do not end up with pollution, overexploitation, and degradation of valuable habitats – on top of the ecosystem devastation wrought by climate change. If we are to avoid this, we must learn from the management of other areas. The Arctic can become the place where we take the time to find good solutions for a rapidly changing system, so we select the right solutions right from the beginning, rather than repeating previous mistakes. The question is which values we choose as the foundation for future development.

The power plant in Longyearbyen, the only one in Norway that burns coal, began generating heat and electricity in 1983 (Photo: Tone J. Sund)

The race for carbon capture and storage

by Ragnhild Rønneberg, former general manager, UNIS CO_2 Lab A/S, special advisor at the Research Council of Norway since 2014

September 2013. No sooner had the new report from the United Nations' Intergovernmental Panel on Climate Change[38] and a book about local changes[39] landed in our laps, before super-typhoon Haiyan made landfall in the Philippines with devastating force. It was an incredible disaster – not least for the Philippine people. Thousands of lives were lost; buildings and infrastructure lay in ruins. An entire world had a rude awakening when the Philippine climate envoy Naderev Sano gave his opening statement at COP19 in November 2013 at the UN Climate Change Conference in Warsaw. With tears in his eyes he implored that we "stop this madness". He and many others are convinced that anthropogenic climate change has a negative effect on the environment and leads to more extreme weather, causing devastation for humans and animals.

"Leaders must act. Time is not on our side," said UN Secretary General Ban Ki-moon when the most recent report was presented November 1, 2014. "Please," he added.

The latest IPCC reports are not encouraging reading. The main conclusions are clear:

- It is 95% certain that humans are responsible for the global warming in the last 50 years
- The temperature will continue to rise, but has not risen as much in the latest 15 years as previously predicted
- Huge volumes of ice have melted in Greenland and at the poles

- Sea level will rise more than previously predicted
- Regions with high precipitation will get even more and arid regions will get even drier
- More storms and extreme weather can be expected

There is no longer any doubt that human activities are changing the planet – we are destroying ourselves. It would be tempting to dwell on all the causes and discuss them in relation to possible countermeasures, but that will not be the theme of this chapter.

Nature whimpers when we do not treat it kindly. This is a fact we can no longer sweep under the rug. We have obligations even here in the chill beauty of the High Arctic, and Svalbard is making its mark on the map where climate research and possible ways solving the CO_2 emission problem are concerned.

The culprit has been identified

Since humans began burning fossil fuels (coal, oil and natural gas) to produce heat and electric power for industry and transport, the amount of carbon dioxide (CO_2) in the atmosphere has increased considerably. CO_2 is the radiatively active gas that contributes most to the greenhouse effect, giving the world a long-lasting fever. The "culprit" goes by the chemical name of carbon dioxide, but one need not be a chemist to understand the possible implications of a rising temperature and the damage it might cause.

In 2013, a new world record was set when atmospheric CO_2 exceeded 400 ppm (parts per million). This measurement was taken at Zeppelin Station in Ny-Ålesund, Svalbard. Thus it is not just in densely populated cities in central Europe, America and Asia that CO_2 levels are high. The number 400 ppm is really nothing more than that: a number, a concentration that experts can relate to. For most of us it is enough to know that this number, this concentration, breaks a record three million years old. That is nothing to be proud of. The trend threatens many species – including humans.

Combustion of fossil fuels such as coal, oil and natural gas has transformed vast stores of fossilized carbon to CO_2, mobilizing it to the biogeochemical cycle in quantities we can no longer handle. The Intergovernmental Panel on Climate (IPCC) states clearly: We must remove CO_2 from the atmosphere. If emissions continue to rise at the present rate, without countermeasures, in just 25–30 years the atmospheric levels of greenhouse gases will be so high that the two-degree target will be unattainable.

To be fair, CO_2 isn't all bad: it is also a crucial component of the atmosphere and provides the carbon required for the life-giving processes of the carbon cycle. Without CO_2, there would be no plants – edible or inedible – and we humans would be in trouble. Against this backdrop it is correct to mention that deforestation is an important cause of increased CO_2 emissions. Plants and trees and other vegetation take up this gas and use it to produce the energy they need to grow and blossom. CO_2 is for plants what oxygen is for humans: we cannot survive without it.

But what happens during combustion of fossil fuel is that the air in many places becomes so thick, so full of CO_2, that it can almost be "cut with a knife". It brings to mind images etched in our consciousness, showing England during the Industrial Revolution at the end of the 18th century and on into the 19th century, when coal was the main source of energy. Granted, the problem then was not CO_2 in the air, but soot and smoke. Nonetheless, fossil fuel combustion increases the amount of CO_2 in the atmosphere.

In 2013, the atmospheric level of CO_2 was 142% of what it was at the beginning of the Industrial Revolution. Many historians consider the Industrial Revolution (about 1770–1870) as the time of greatest technological change in human history since the dawn of agriculture in ancient Mesopotamia 10,000 years ago. Technological development is normally seen as good, but we now realize that the Industrial Revolution introduced practices that are no longer sustainable.

Black smog full of coal dust and exhaust was considered unhealthy in the 20th century, and it has not become any safer with the passage of time. In parallel with increasing emissions of CO_2, we are gaining insight into the risks posed by high levels of CO_2 in the atmosphere and the oceans. We know a lot about the negative consequences of increased CO_2 on land and sea, and its impact on living species. These problems will not resolve themselves over time, and the CO_2 does not disappear on its own, so we need to take measures to cut our O_2 emissions even as we adapt to a changing climate.

We must start another revolution, a revolution focused on new energy sources, energy production, and energy consumption. We can no longer expect – or hope – that nature will adapt to us. We must adapt to nature. That means we must toe the line and act so we do not ruin everything.

Svalbard – a protected wilderness where coal is king

Welcome to Svalbard – a region of scenic beauty, exotic surroundings – and extreme sensitivity to climate change. All but 2% of Svalbard's land area is untamed nature; 98% is wilderness, of which 65% is under protection to preserve the natural environment, the beautiful landscape, and the cultural heritage sites (all signs of human activity prior to 1947 are protected). For comparison, only 12% of the Norwegian mainland is under protection.

Around 60% of the land in Svalbard is covered by glaciers, large and small. Some of them attract droves of tourists in the summer. Tourists often climb to the best vantage points atop cruise ships, or approach the glacier in smaller vessels, hoping to get close enough to capture all the drama. Beautiful turquoise walls of ice tower above them, ice from millions of years ago stored in yearly strata like the layers of a cake. The sight of a calving glacier is both beautiful and thrilling. Flying low over a glacier and peering deep into its crevasses – where blue-green chasms plunge to unfathomable depths – evokes a fascination that is equal parts drama and danger. And even as the glacier calves, water trickles beneath it – trickles, flows, gushes – growing in volume and speed, and one begins to understand that this signals something about the glacier's health. The more the ice melts and the more the glaciers calve, the more the sea level rises, and the more freshwater is introduced into the salty oceans. These changes affect both weather and wind and have implications for the entire ecosystem.

This chapter tells the story of a project for the environment run by the University Centre in Svalbard (UNIS): the UNIS CO_2 laboratory project. Right in the middle of the jewel of the Arctic. You have arrived in Svalbard after a 2012-kilometer flight from Oslo/Gardermoen, emitting an extra 208 kilos of CO_2 per passenger. Have you by any chance noticed that SAS gives their customers an opportunity to compensate their share of the CO_2 emitted during the flight so the journey becomes CarbonNeutral®? For a mere 2.27 EUR you can buy yourself a clean conscience. But few travelers do so.

Don't take it personally. You are certainly not the only one who emits CO_2 from aircraft and other means of transport. Nor are you the only one who doesn't know what to do about the amount of CO_2 you emit – today or in the future.

But we all agree that the CO_2 (and other greenhouse gases) that we emit makes the blanket around the earth a bit warmer; and we have reason to worry about the future if we continue to emit CO_2.

There are good indications that we are starting to care about our climate, the environment, and the changes we see nearby and far away – aren't we? There's a connection here, isn't there?

Coal is the king of all energy sources – in Svalbard as in the rest of the world. Use of coal is increasing in many countries, not least in Asian countries such as China and India. China alone now consumes about half the coal used in the world, whereas coal consumption is falling in the United States, largely because of increased access to shale gas. In 2012, total emissions of greenhouse gases decreased in the European Union, but emissions specifically from coal increased by 3%. In Svalbard and Longyearbyen, coal is the only alternative, at least in the foreseeable future.

Svalbard's communities are based on coal – and have been since the early 1900s. Coal still runs Svalbard, not just in Longyearbyen and Svea, but also in Barentsburg. Coal mining is one of the foundations on which Svalbard's economy stands, and nothing suggests that mining will cease even if the local communities find other sources of employment; Norwegian presence in the archipelago is far too geopolitically important. What energy sources shall we use for residents and visitors in this region – one of the most important thermometers for measuring the earth's health? If not coal, *what*? As long as the energy source remains unclear, we are ignoring the elephant in the room.

For far too long, we ignored the problem of increasing CO_2 emissions and hoped that nature would solve our problems, or that others would shoulder more responsibility than we ourselves can carry. All changes come at a cost, changes in habits too, but it's nice if someone else picks up the tab. It's the same with CO_2 – a contest of moves and countermoves, frequently ending in deadlock. Coal, oil and natural gas block development of energy from the sun, the wind and other renewable resources – in Svalbard, too.

When people come to Svalbard, it is usually not because they want to see the melting ice with their own eyes, but some visit with precisely that objective. The paradox is that they cannot get here without huge emissions of CO_2. Only a handful travel with zero-emission transport, such as sailboats. What this boils down to is that travel to, from and within Svalbard contributes to CO_2 emissions, from planes, cruise ships, cars, buses, and snowmobiles.

It is completely unrealistic to expect the demand for and use of these polluting means of transport to change within a generation or two – neither here nor elsewhere. Emission of CO_2 from cars, busses and snowmobiles in Svalbard is negligible in a global context. Besides, the main sources of CO_2 emissions in Svalbard are undoubtedly air travel and – not least – the local power plant in Longyearbyen, where we have Norway's only coal-fired power plant. It is worth mentioning that there is also a small power plant in Barentsburg, the Russian settlement west of Longyearbyen. The CO_2 emissions from these two plants are modest, both in terms of total amount and in comparison with just about any other power plant that runs on fossil fuel. Still, the plants exist and are constant reminders of the climate issue that is so easily observed in this region. The power plant in Longyearbyen isn't even considered a substantial enough emission source to warrant research aimed at reducing emission of greenhouse gases.

What can we do that will not be purely symbolic, but actually solve a local problem and simultaneously advance knowledge and technology in ways that other nations might put to good use? This tiny community has everything required to show that carbon can be handled in an environmentally responsible manner.

A drop in the ocean?

The power plant that provides light and warmth to the buildings of Longyearbyen emits about 60,000 tons of CO_2 per year. This is a drop in the ocean compared with other emission sources in Norway, such as Norcem's cement factory at Brevik, the Yara fertilizer plant in Porsgrunn, the Ironman iron processing plant at Tjeldbergodden, and Industrikraft Møre at Elnesvågen (a power plant that emits about 1.8 million tons of CO_2 equivalents annually). Yet these are tiny emission sources compared to coal-fired power plants and energy-consuming industries in other parts of the world.

Even if we include all of Norway's emission sources, the total is a drop in the ocean compared to emissions in other countries. From that perspective Norway is a rather modest direct emitter of CO_2. It is best to forget (at least in these calculations) that Norway exports oil and natural gas that give rise to formidable CO_2 emissions in other parts of the world. As proud world champions in many different disciplines – including knowledge about CO_2 and how to handle it – we should take responsibility and make a difference that amounts to more than a drop in the ocean. We have a moral obligation to do something to neutralize our CO_2 footprint in the years to come.

Norway's domestic emissions have changed over time, though they are still far too high. We mustn't moralize, but allow the numbers to speak for themselves. In an ethical debate about climate change and the role of CO_2, we must acknowledge the elephant in the room: the petroleum industry, other heavy industries and the transport sector are the main puppetmasters behind CO_2 emissions.

The oil/gas and transport sectors are definitely going in the wrong direction: emissions increase with every passing year. On the other hand, CO_2 emissions have decreased in agriculture and in the heating of homes and businesses.

We need not delve deep into data from Statistics Norway to find numbers that take our breath away. Per capita emissions in Svalbard total more than 68 tons of CO_2-equivalents (i.e. carbon dioxide (CO_2), methane (CH_4) and nitrogen oxides (NO_x) combined). This is because of the coal mining and local coal-based energy production. CO_2 emissions from ships and planes are not included in those 68 tons. Regardless, Svalbard residents top the list, using more than the citizens of Qatar (55 tons per person per year in 2011) and consuming more than ten times the global average.

Per inhabitant, Norwegian CO_2 emissions are on par with those of western European nations. Norway's emissions per capita are less than half those in the United States, but substantially higher than those in developing countries (Statistics Norway, 2012).

Of course calculations per capita never tell the whole truth. Total emissions worldwide add up to about 34 gigatons – a number and a quantity few of us can relate to. China leads the "emissions race" with about 9.7 gigatons of CO_2. In second place comes the United States with about half as much, and India and Russia are far behind. Norway's domestic emission of CO_2 is about 44 megatons per year, or 9 tons per capita (data from 2013). If we include CO_2 from combustion of exported Norwegian oil and gas, emissions jump to 500 megatons per year, or 110 tons per capita.

A report published in 2009 (by the Norwegian Environment Agency, NILU – the Norwegian Institute for Air Research, and UNIS) showed that growth in tourism to Svalbard contributes to increased emissions. This is no surprise, but is nonetheless a dilemma. We want people to visit Svalbard, experience the pristine, vulnerable environment, and gain insight into the climate and environmental challenges – which are so easy to observe here – but it comes at a price. The more people who come to Svalbard, the greater the emissions.

According to the report, soot emissions increased by 56% over a seven-year period, NO_x by over 50%, and CO_2 by 30%. The study concludes that most of the emissions come from coal mining and ships, particularly cruise ships. Thus tourism is not an entirely good thing – seen from the perspective of CO_2 release – and tourism is on the increase. Nonetheless, total emissions in Svalbard are negligible both in relation to other places in Norway and – not least – in the rest of the world.

But that doesn't mean Svalbard should be exempt. After all, we pride ourselves on having "the world's best preserved wilderness area". We are obliged to live up to that; we should set a good example. At issue here is emission of 126,183 tons of CO_2-equivalents from Svalbard (in 2011), versus 14,356,952 tons from all the land-based industries in mainland Norway. Svalbard's emissions are indeed a drop in the ocean, but nonetheless, they don't look good.

We have no oil or gas activities in Svalbard, and very little transportation, but we can't dodge responsibility. We mine and burn coal that contributes to CO_2 emissions here and abroad. That cannot be denied.

The dilemma is complex from the perspective of Svalbard, where the local community light and heat their homes with coal. Norway discourages increased use of coal in the rest of the world, though we export coal to other countries. In addition, Svalbard has a growing influx of tourists who contribute to CO_2 emissions through their travel, their activities, and their mere presence in the archipelago. How will this affect the CO_2 balance sheets?

There is one small consolation: the coal from Svalbard is cleaner than most other coal, for instance brown coal from Poland. Svalbard's coal has high energy content and emits less CO_2 per unit of heat produced. Setting a good example is important in the CO_2 game. In addition, coal mining helps ensure Norwegian presence in Svalbard, making it a useful card when the High North is the prize.

Coal – threat or opportunity?

If we are to make wise moves to prevent future negative trends, we first need to understand the components of global warming. At the same time, it is imperative to develop a portfolio of activities that can bear fruit both in the short and the long term. We need many different activities – and they are not mutually exclusive:

- reduce emissions of CO_2
- develop green/renewable energy sources
- study and develop technology for capture and storage of CO_2

According to the International Energy Agency, carbon capture and storage (CCS) can yield 20% of the CO_2 emission cuts required before 2050. Through the United Nations, the world has set the goal of limiting the rise in global average temperatures to 2°C above pre-Industrial times. But to achieve that goal, emissons need to start falling to low levels before 2050.

To avoid global warming exceeding 2°C, about 3,000 CCS systems must be installed and fully operational before 2020. So when must we start building those systems? Yesterday!

The Energy Agency estimates that about 70% of our future energy will come from fossil fuel – which underlines the importance of implementing CCS. Combustion of coal is one of the worst culprits in this context, contributing about 50% of the world's CO_2 emissions. CCS facilities on coal-fired power plants (old and new) would therefore have a huge impact, reducing CO_2 emissions both significantly and immediately.

Coal has been the main source of energy in Svalbard since 1905, when the American industrialist John Munro Longyear founded the Arctic Coal Company. In 1916, the company was bought by the Norwegian mining company Store Norske Spitsbergen Kulkompani, which currently operates three coal mines in Svalbard: *Svea Nord* (60 km south of Longyearbyen), *Gruve 7* (in Adventdalen on the outskirts of Longyearbyen), and the recently opened mine *Lunckefjell* near Svea Nord. Calculations show that the coal in Lunckefjell will yield CO_2 emissions of about 5.5 million tons per year when used in the European energy and industry sector.

Gruve 7 provides coal for the local power plant in Longyearbyen, which generates electricity and heat for the entire town. As Norway's only coal-fired power plant, it consumes about 25,000 tons of coal and emits 60,000 tons of CO_2 annually.

The Longyearbyen CO_2 Lab was inaugurated in 2007 with the vision of creating *CO_2-neutral Svalbard*. Within this vision lay an aspiration to capture CO_2 emitted from the power plant and store it in a responsible way, as well as reducing emissions from cars, snowmobiles and other vehicles.

The plan was to capture carbon emitted in the form of CO_2 from the power plant and store it in the ground. There was also a broader ambition: through targeted research and

UNIS CO_2 Lab in Adventdalen outside Longyearbyen (Photo: Ragnhild Rønneberg)

development, Longyearbyen could become a showcase community – a community that took responsibility for its CO_2 emissions and contributed locally toward reducing global atmospheric CO_2 levels.

Recognizing the need to smooth the way for green energy – even in the Arctic – it is necessary to invest in research, technology and knowledge. In addition, incentives must be brought into effect to make environmentally friendly, energy-efficient options more attractive, even in a small community such as Svalbard.

Greenhouse gases / climate change culprits

Many factors influence the climate: an increased CO_2 level is just one such factor. Methane (CH_4) is another – a greenhouse gas that is 25 times more efficient than CO_2, per carbon atom. Ruminants are often listed among the worst perpetrators of climate change because they release methane. When they digest their food (plants with high cellulose content), extracting energy and nutrients, large amounts of gas are released – both CO_2 and hydrogen

(H_2). Microorganisms in the stomach convert the hydrogen to methane, which the animals then eliminate by burping or farting. Ruminants contribute about 15% of total greenhouse gas emissions, but the cows themselves don't seem to be very concerned.

In Svalbard – in the Arctic in general – methane release isn't related to ruminants, but to a source in the ground itself, more specifically permafrost. As long as the permafrost remains frozen, most of the methane stays put, but the moment the ground thaws, methane escapes in quantities that threaten the entire planet. But that's another story. Here the focus will remain on CO_2 emissions and our possibilities of removing CO_2 from the atmosphere – or even better, never letting it reach the atmosphere in the first place.

Where do our CO_2 emissions go?

From 1750 to today, there have been two main sources of CO_2. About 70% of the CO_2 emitted comes from fossil fuel and production of cement; the rest comes from deforestation and changes in land use. Where has this CO_2 gone? It doesn't disappear simply by magic or because we don't like it. Well, nature itself takes care of our emissions to some degree. Oceans absorb about 30%; trees, herbs and other vegetation take up another 30%. The rest, just over 40%, remains in the atmosphere. As we have known for many years, the amount of CO_2 is increasing – both in the oceans and in the atmosphere. Now we are no longer talking about proportions and percentages, but about absolute numbers, the actual amounts. That is what makes the difference – huge amounts of CO_2 that we don't know how to handle, but that we *do* know will cause unwanted climate change. We have already reached the point where urgent steps must be taken to ensure that the amounts of CO_2 do not to continue to rise, but actually decrease, so the equilibrium between production and consumption can be re-established.

Nature can adapt to a lot, but over time it will be too much, and CO_2 emitted in the future will remain in the atmosphere longer. If that happens, it will reinforce climate change; the need to cut emissions and to adapt to a changing climate will be even greater. The longer we wait, the more difficult it will become. It may even become impossible…

How often do you take your car in for service? Nobody waits until their car is a wreck, and then expects it to be as good as new when it leaves the repair shop. The same thing goes for our environment. Measures to curb CO_2 emissions must be taken before it is too late.

How can we catch the culprit?

Capturing and storing CO_2 is necessary – not as the *only* measure, but as one of the most important measures aimed to reduce CO_2 levels in the atmosphere. Carbon capture and storage involves extracting CO_2 from a flow of gas (exhaust) and storing it permanently in geological reservoirs underground. Carbon capture and storage (CCS) has become a standard international term, denoting both the process itself and the research field as a whole. Technological solutions for large-scale CO_2 capture are already commercially available and well developed.

Although CO_2 has previously been injected into geological formations for various reasons, the concept of long-term storage of CO_2 is relatively untried. But in several places, CO_2 capture technology is advancing rapidly. Other links in the CCS chain have also reached the point where they are ready for implementation. It is not unreasonable to hope for at least a minor technological revolution in this field quite soon. The activities that should be first in line for CO_2 extraction are power plants, cement factories, and other factories that run on fossil fuel.

CCS is not particularly complicated, but nonetheless presents several chemical, physical and technical problems for extraction/capture, transport and storage. Here, in brief, are the issues:

1. ***Capture/purification***: CO_2 must first be separated from the other gases (nitrogen, oxygen, sulfur compounds), water vapor, particles and soot in the exhaust from combustion of coal, oil or natural gas. The purification techniques that are currently available capture 80–90% of the CO_2, and consume extra energy. This means CO_2 cleanup increases costs. In addition, these processes require use of chemicals that are not only considered detrimental to health, but also make the entire process more expensive.

2. ***Transport***: Once CO_2 is extracted from exhaust, it must be compressed, transferred to a special type of container (made of stainless steel), and transported to a storage site by pipeline or in specially designed vessels. Whatever the method of transport, all the equipment involved must be highly resistant to corrosion, because compressed CO_2 becomes caustic in presence of water. This is another costly step.

3. ***Storage***: Getting rid of the captured, compressed CO_2 once and for all requires a suitable disposal site. But how do we dump this "garbage bag" so it will not be hacked

to pieces by scavenging crows? Clearly anything remotely like a compost heap in the backyard or an open-air junkyard would be out of the question. We must locate and verify specific geological formations capable of receiving these huge quantities of CO_2. What we are looking for is porous, fissured rock. The most suitable geological formation is porous sandstone – and it must also be capped by a layer of impermeable rock. An empty or abandoned oil or gas reservoir could also be used for CO_2 storage. Drilling wells for injection and observation is costly. Moreover, the wells must be designed and constructed to tolerate "aggressive" CO_2 – another costly requirement.

These considerations raise two new issues or challenges:

1. costs
2. storage time

Presumably, technological advances over time will reduce costs at all stages of the process from CO_2 capture (be it from coal- or gas-burning power plants or heavy industry) to safe CO_2 storage on land or at sea. It is always the first few facilities that cost most to build. Projected costs for capture range from 150 to 1500 NOK (20–200 EUR) per ton of CO_2. The variation in cost estimates reflects geographical conditions, requirements for infrastructure and technology, and whether the CO_2 to be captured comes from combustion of brown coal, higher grade coal or natural gas. Costs vary considerably, and the estimates are uncertain. Against the backdrop of the current cost of emitting CO_2 (as little as 5 EUR per ton, according to the European emissions quota price system) there is little or no incentive to start implementing capture and storage of CO_2. What is the solution? Higher prices for emission quotas? Lower technology costs? "Yes, please. Both!" Technological research and development is at any rate ongoing in many countries – not only in Norway.

The high costs themselves raise an ethical question. Is it appropriate to use however much money it takes to solve problems *now* – to prevent further deterioration *now*? Or should we postpone problem-solving in the hope that the solutions will be cheaper in the future, or that we can adapt to climate change, and all the problems it may potentially cause for living organisms on Earth?

If we disregard the issue of cost, can we just roll up our sleeves, get going and hope for the best? No, we can't. At present nobody can guarantee that the CO_2 pumped down into

Geological conditions in Svalbard are perfect for storage of CO_2 (Photo: UNIS)

a reservoir will stay there "forever". This problem is at the core of CCS. Nobody can take responsibility for what is pumped into the ground. Some countries, such as Germany and Denmark, actually prohibit storage of CO_2 on land, and therefore expect the "captured" CO_2 to be transported away. But is it *fair play* to dump your trash on your neighbor's property? This is one of the challenges of CCS, not least within Europe.

How can we keep the culprit imprisoned, and for how long?

Longyearbyen CO_2 Lab started from research on storage conditions for CO_2: storage underground on land – not storage at sea, on the seabed several hundred meters below the surface.

To explore the conditions for potential future storage of CO_2 captured locally in Svalbard, a whole series of studies were carried out: boreholes were drilled, cores were examined, and samples were taken from outcrops of rocks suitable for storage. In addition, researchers made computer models and projected scenarios. (Modeling is a type of statistical analysis in which large sets of data are evaluated in search of mathematical relationships between different variables/parameters and measurements.)

Six wells were drilled, ranging in depth from a few hundred meters to nearly one kilometer. These wells are at the UNIS CO_2 Well Park in Adventdalen, near the old Auroral observatory.

A series of test injections were conducted in some of the wells, while others were fitted with sensors and instruments and served as observation wells. These studies made it possible to map a reservoir in sandstone formations 670–970 meters underground.

Numerous injection tests performed with various amounts of water at different pressure, combined with modeling results, suggest that this reservoir can hold CO_2 corresponding to at least 20 years' emissions from the power plant that is currently in operation. The total amount emitted is relatively modest, as mentioned before, but it is enough to establish a research platform to study geological CO_2 storage on land.

So we have identified a "storeroom" – a reservoir, as it is called in the technical jargon. What else do we need to check before pumping any CO_2 into the ground? Obviously we must make sure the reservoir doesn't have a leaky roof – and we were also able to verify that. The reservoir in Adventdalen is covered by a dense shale roof no less than 200 meters thick, with another overburden on top of that. Above these is another lid of about 100 meters of permafrost. It might be compared to having three protective cushions around a champagne bottle. This bodes well!

So far, the CO_2 Lab has only tested the "storeroom" with water – safe and unproblematic in every way. But when we start injecting the real thing, CO_2 itself, it is another matter entirely. One concern is that cement and pipes in the installation will be of inadequate quality and will disintegrate in contact with corrosive CO_2. Another question is how CO_2 molecules will behave in the sandstone pores and fractures, which already contain water and other substances that could react with them. Will the CO_2 molecules react with components in the rock and become part of the rock itself, so they never come out again? Will the CO_2 dissolve in the water, or be in liquid or gas form so that it might leak out at some time? Where would liquid or gaseous CO_2 go? Would it rise up through the boreholes and pipes, or spread toward Tempelfjorden, where the rocks of the reservoir appear in outcrops, or emerge through craters on the seabed in Isfjorden? Such craters, called pockmarks, already exist. They are caused by eruptions of gas or fluids that have flowed laterally through the sediments, and could serve as conduits for eruption or flow of gas triggered by injection of CO_2 on land.

When, where, and how CO_2 might leak is still a mystery. Therefore, much research is focused on understanding potential geological changes on land and in the seabed when large amounts of water are injected into Adventdalen – and these studies must be carried out and evaluated before any CO_2 is injected. We have determined that the pressure in the reservoir is quite low. This means that the surrounding rocks will function like a vacuum, sucking up and holding fluids and gases at depth and consequently reducing or eliminating the risk of leakage toward the surface.

What we know for certain is that the geology of the reservoir in Adventdalen is suitable for storage of an amount of CO_2 corresponding to at least 20 years' emissions from the power plant. In addition, we know that the world needs pilot projects on carbon storage – and Longyearbyen is well placed for initiatives to study land-based CO_2 storage.

In other contexts and at other sites, undersea CO_2 storage will be more appropriate, not least in oil and natural gas fields that have already been emptied. It is critically important to study the various opportunities these fields and the associated sedimentary formations have to offer. Unfortunately, techniques that work well at one site can never simply be applied to another. Nature is too diverse and too unpredictable. But knowledge gleaned about methods and instrumentation at one site can be transferred to others. That is good to keep in mind when Norway's CCS initiative needs to find new paths.

Many questions remain unanswered, and much painstaking research remains before any large volumes of CO_2 are injected into the ground in Adventdalen.

Against this backdrop, we plan to inject small amounts of CO_2 (up to 100,000 tons over a period of up to ten years), and monitor closely to detect any leaks. This will require use of highly specialized, exquisitely sensitive equipment. In this way we will close a few more gaps in our knowledge of what constitutes "safe carbon storage on land".

The goal of the UNIS CO_2 Lab is to demonstrate safe storage of CO_2 (preferably produced by and captured from the local power plant), but also use this arena to demonstrate and refine knowledge about storage in this type of rock formation, which can be found elsewhere in the world (e.g. South Africa). The establishment of a full CCS facility would fulfill

Longyearbyen's aspiration to become "a green showcase" – a small community that takes responsibility for its own emissions.

Is it safe?

A certain degree of skepticism is healthy. The unknown often evokes both faint anxiety and a bit of opposition. It is the same for storage of CO_2. It is only natural that people wonder how safe it is to store CO_2 in places where no CO_2 has been stored previously. The NIMBY syndrome (for "not in my backyard") is strong also when it comes to CO_2 storage.

We are learning as we go – this research is literally groundbreaking. New knowledge is generated when reservoirs that contain gases are monitored continuously. Obviously gas can leak out when boreholes are drilled in the earth's crust, and when cracks form. Gas from natural sources can also leak out through pre-existing cracks and faults; these leaks are totally unrelated to injection of CO_2.

For research purposes, it is important to provoke leakage from specific storage locations both to understand the leakage phenomenon itself, and to develop instruments sensitive enough to measure small amounts of leaking CO_2. It is necessary that every aspect of CO_2 storage is based on technological and scientific understanding of natural systems – with and without leakage.

Knowledge gleaned from natural analogs is also extremely valuable. Nature demonstrates that under certain geological conditions, gas can be kept in place for thousands and even millions of years. This makes it important to know about various geological formations – know whether they contained oil or gas in eons past, or might potentially store gas in the future. All geological formations differ.

The gas CO_2 is a natural part of the atmosphere and is necessary for the survival of all plants. It is only at very high concentrations CO_2 is dangerous to humans. CO_2 is a colorless, odorless gas; it is heavier than air and displaces oxygen from human red blood cells – which is why it disrupts normal body functions when it exceeds a certain concentration. There is no risk until about 5% of the air we breathe is CO_2. Levels that high can lead to headache, dizziness, and nausea – but the symptoms disappear if one moves away from the CO_2 source. CO_2 levels in the air in our lungs regulate our breathing – we want to inhale oxygen and exhale CO_2. (CO_2 is formed in our bodies as part of our normal metabolism.) If our lungs contain abnormally high concentrations of CO_2, we breathe faster. In air containing 2% CO_2, we breathe 50% faster than normal. We don't feel at our best, but we can get by.

At concentrations of 15–16% the situation is more serious, and protracted exposure to CO_2 can even be fatal. Obviously, those who work in CO_2 capture, transport and storage facilities might be at risk if something went wrong, but people in general would not be threatened by leakages from man-made wells or natural fissures, because any CO_2 above ground in open terrain would soon disperse. Even the slightest wind would suffice. In Adventdalen, the air is constantly in motion – so it would be a safe place to inject small amounts of CO_2 into the ground for research purposes. Nonetheless, in the interest of safety, such testing must be preceded by a thorough risk assessment in line with requirements from the Norwegian authorities.

We've done it before – and we're still doing it

Norway has already come far with capture and storage: two Norwegian offshore gas fields, Sleipner and Snøhvit (Snow White), are often cited as good examples of how to handle CO_2 – a sort of "look to Norway" slogan in the context of CCS.

Statoil started separating CO_2 from the natural gas pumped up at Sleipner in 1996, and since then, about a million tons of CO_2 have been re-injected annually. At Sleipner, CO_2-rich natural gas is transported to the Sleipner West operation, where CO_2 is separated from the natural gas in a capture facility. The extracted CO_2 is then re-injected and stored in a saltwater reservoir, the Utsira Formation, a geological formation consisting of porous sandstone about 800 meters into the seabed. It was not environmental concerns that originally prompted CO_2 extraction at Sleipner, but the fact that the high CO_2 content made the natural gas difficult to sell at a profit.

Snøhvit, which opened in 2008, is another proud example of Norwegian carbon capture and storage: approximately 0.7 megatons of CO_2 are pumped into the Tubåen formation under the Barents Sea after being extracted from natural gas from the Snøhvit field. Also at the Gudrun field, which became operational in August 2014, CO_2 will be extracted from the natural gas and stored in the Utsira formation. But even these good examples constitute less than 1% of the volume that must be stored on a global basis.

It is beyond doubt that incredible amounts of new knowledge and technology have developed over the years concerning safe, secure carbon capture and storage on the seabed. Norway's Petroleum Safety Authority monitors the work and has continuous follow-up. The Authority's eyes and ears and nose on the seabed are sophisticated instruments that measure, monitor, and "sniff out" any leaks. So far everything looks fine. But it is different when land-based storage is planned in places without natural "open storerooms" that have been emptied of oil or natural gas – particularly in places where people live and work. Places like Potsdam (Germany), Lacq (France), Hontomin (Spain), Sicily (Italy), and the tiny village of Longyearbyen. These are just a few of the places where pilot facilities for carbon capture and storage are planned or already in operation.

Dreaming about the moon is easy. Getting there is much more difficult and risky. It requires many years of methodical preparation. "That's one small step for man, one giant leap for mankind," said Neil Armstrong when his left foot touched the surface of the moon in July 1969, early in the morning Norwegian time. Armstrong and Aldrin spent 2½ hours exploring the moon's surface before returning to Earth. Nothing was ever the same after that – not for the astronauts, and not for the rest of us on the planet.

More than forty years later "moon landing" would become a political concept in Norway[40]; it stood for soaring ambitions to save the world using Norwegian technological know-how to capture CO_2 from gas. We will draw a veil over how the nation's attempted "moon landing" in the form of full-scale CO_2 capture at the land-based gas-burning power plant in Mongstad ended in fiasco in 2013. It is only fair to mention that the attempt led to significant technological advances and generated important new knowledge that will be extremely valuable in future development of CCS – be they here or there, large or small. If the costs soared right along with the ambitions, that is another matter entirely. And that cleaning gas exhaust differs from cleaning coal exhaust is also another matter.

Another question also remains. Supposing we *have* mastered carbon capture, where can we store our catch? If we look only at options within Norway, there is room for about 5.5 gigatons of CO_2 in the Norwegian Sea, according to the Norwegian Petroleum Directorate's new storage atlas. That is more than one hundred times as much CO_2 as was released in all of Norway last year. Room for everyone! For the time being, large-scale storage appears to be the key to unlocking CCS on a global scale.

Why Svalbard?

A CO_2 project in Svalbard has the advantage that it "fits" in a smaller scale, and for the purposes of research and development. The area is geologically suitable; many people (including policymakers) visit Svalbard in person "to see the melting ice"; and the international university, local businesses and stakeholders in Longyearbyen offer much relevant expertise. Everything in one place, in other words!

Clearly a project of that nature has twice the persuasive power when it is situated in the Arctic – the region where climate change will undoubtedly become visible soonest and proceed fastest.

This has always been a high-profile project, reaching out both to the general population and to scientists and politicians around the world. Deliberate efforts have been made to facilitate information transfer and education, both at UNIS itself and – not least – at the site where the actual drilling and testing activities take place. In this way, the project has highlighted the opportunities CCS has to offer, along with the important roles played by research, development and education.

The CO_2 project also aspires to establish an Innovation and Knowledge Center that will demonstrate and provide information about a wide range of energy sources – from solar and wind-generated power, to coal-based and geothermal energy, to name just a few. Once again, the Arctic – particularly Longyearbyen – is a suitable and convenient location for this type of outreach and information transfer both about making fossil fuels sustainable and about emerging "green" technologies.

UNIS CO_2 Lab was in the running for some time, with our strong focus on CO_2 storage – at least in terms of research and testing in Svalbard. But high costs and political vacillation hampered what could have been the world's first site for greener use of coal and local management of CO_2. In the end, the Canadians beat us to the finish line. In October 2014 the world's first full-scale carbon capture facility for a coal-fired power plant opened in Boundary Dam, Canada. It came at a price. For the tidy sum of $1.24 billion (plus $115 million over budget a full year before start-up) this will undoubtedly be an important model site, demonstrating how to handle CO_2 emissions from power plants run on coal. Like Longyearbyen, Boundary Dam has no local energy source other than coal. Coal power is the only alternative – but now there is a capture facility. It will be exciting to follow this project over time, both in terms of technology for CO_2 capture and transport and, not least, CO_2 storage conditions. A lot is at stake here. If everything goes according to the Canadian plan, Boundary Dam will be a good investment for the rest of the coal-burning world. This reveals something about the gamble that is CCS. There is a lot of money at stake. Captured and compressed CO_2 can be injected into sandstone formations to increase yield in oil and gas fields. This means it can generate income – and that is also part of the gamble. At Boundary Dam they plan to capture and sell not only CO_2, but also sulfur dioxide (SO_2). Such "fringe benefits" are an important factor in the overall economic feasibility of CCS. It's like a horserace – everyone wants to bet on a winner.

A single sentence says it all

"If humanity is unable to limit our emissions of CO_2 from fossil fuel, we may face famine, life-threatening sea level, impoverished biodiversity, desertification, millions of climate refugees, submersion of countless islands, and uncontrollable spread of diseases." This was the message in the daily papers when the 5th IPCC report was presented in March 2014.

I was probably not the only person who gasped. Others wondered "How long do we have?" That question has no precise answer. A message presented this dramatically can easily lose some of its credibility – particularly when we cannot observe the effects at first hand.

Perhaps the headlines overdramatized (as media messages often do). But nature itself is dramatic. That is why we find it so fascinating – in Svalbard, South Africa, and the Amazon; in Alaska, China and at the South Pole; wherever we happen to be on our beautiful planet. Either we team up with nature or we don't. The CCS race has already started – and we try to catch up with many fine words, some new technology, and lots of knowledge. But there is also inactivity, resistance and reluctance, because it is painful to change our way of life, our behavior, our consumption patterns. The *planet* will survive. The question is not who will prevail: nature will triumph over us. The question is *how much* we will lose. Perhaps life itself, if you can bear to think the thought to its full conclusion. Few now doubt that the situation is serious.

The 5th IPCC report, presented March 31, 2014, reiterates what we already know. Climate is changing all over the world, and the situation is serious. Nature is no longer what it used to be – not even in Norway or Svalbard. This should nudge us onto better paths; make us pause and think about how our own consumption contributes to CO_2 emissions. Think – and ideally also do something about it!

Individual initiatives matter in the scheme of things

Sometimes a millimeter is enough – enough to initiate a movement toward a more climate-friendly future. If we all do a little, together we can achieve a lot. There is still truth in the notion presented by Norway's former prime minister Gro Harlem Brundtland: *everything is connected to everything else*. Far-sighted investment in the future also requires small-scale initiatives from local communities and individuals. The choices made by each and every one of us (myself included) can actually make a difference overall. Useful strategies include:

- Thoughtful use of energy at home, at work and in leisure time. Use less energy overall – use less energy from fossil fuel – use your head!
- Avoid using transportation systems that run on petroleum-based resources
- Buy less stuff. What about a movement from "3 for the price of 2" to "1 for the price of 1" – or better yet, "Buy second-hand".
- Eat less meat and foodstuffs that cost more energy to produce than they contain in fresh or frozen form.
- Read books that list little things you can do to help the environment. Regardless of whether you intend to change your lifestyle dramatically or just make minor adjustments, you will find suggestions that fit your everyday life. Better, healthier, cheaper!

What are we racing toward?

In the general debate about climate and CO_2 emissions, one aspect is seldom discussed. The larger the number of people who leave poverty behind, the greater the need for energy. This is an upward spiral. Sooner or later there will not be enough energy to go around, neither light nor heat nor energy to produce goods and services. We are getting closer to that point now. Increased need for energy will increase emissions of CO_2 – until enough renewable energy becomes available.

Those who produce fossil energy repeat their mantra: "We mustn't deny populations in other parts of the world access to energy or reject their aspirations for growth and development to the prosperity we ourselves enjoy." No, we mustn't. Some beg to differ and reply "There is no lack of energy. There are just too many people on earth for everyone to have the same prosperity." Two different views of the climate challenge – two widely disparate views of humanity.

The IPCC reports presented in 2014 are gloomy reading. Even as CO_2 emissions continue to increase, the reports indicate that the future holds flooding, drought, conflicts and economic difficulties. People in coastal areas of Asia are among those who will be hardest hit by global warming. More rain is expected in places where rainfall is already high, whereas many arid regions will probably grow even drier. This future looks bleak.

But we are not the ones who will suffer most under a new climate regime – the suffering will be greatest among those who are already the world's most destitute. Basic resources and food production will change dramatically. Too much water in some places and too little in others will lead to floods and drought – and we must get used to it. This obviously affects everyone's life and health. The race will be for water and a reliable food supply. This reaches far beyond local or national competitions about CO_2. One thing is certain: we can get used to a lot, but the CO_2 level must come down – fast. There is reason to fear that not much will happen with CO_2 emissions until strong incentives are in place for both individual citizens and commercial interests, with the entire portfolio of state-owned business interests in the driver's seat. But we can hope that an ethical debate about how we envision our planet and its future will bring us to our senses.

Any major undertaking involves many steps – and where reducing CO_2 emissions is concerned, several of those steps can and should be taken simultaneously:

1. An assortment of measures to *cut emissions* – at home and elsewhere
2. Accelerated development and increased use of *renewable energy sources*
3. Emissions that cannot be curtailed or changed must be subject to *CO_2 capture and storage*

The IPCC has made clear suggestions about what measures we should take – and we should take them. We have recognized both the culprit and the puppetmasters.

We want a good world for ourselves and our descendants. We can achieve it if we try. The road to a low-emission society is passable, and it need not even require a tedious change of course. Not in Svalbard either – though the road is slightly different. Even Svalbard has enough energy from sun, wind, waves and geothermal sources. Coal must not be allowed to block the road for these resources. All energy sources must work together.

But the opposite is happening. To reiterate, more and more countries are making themselves dependent on coal. In China and India, coal is king. Australia is tagging along and wants to increase coal production by nearly 50% from 2011 to 2035. In Germany, solar power is overshadowed by coal. It is true that input of energy based on renewable resources is increasing in Germany, but coal-based power production is also increasing. This is because the price structure for coal and natural gas makes "dirty" coal cheaper than "clean" natural gas, and solar power is even more expensive. Combined with very low costs for CO_2 emissions in Europe (Norway excepted), this leaves little room to maneuver. Unless the choices are based on a different set of values. The global CO_2 race has been underway for some time; many countries are going back on promises to reduce their CO_2 emissions. Some of them (e.g. Germany and Denmark) prohibit storage of CO_2 on their own soil, while others point to the formidable CO_2 storage potential in the empty oil and gas fields at the bottom of the North Sea.

There are indications that renewable energy alone will not suffice to meet the demand on the world market, but that it is essential to implement carbon capture and storage – not least as a means of making coal cleaner. This brings us back to Svalbard, where the Longyearbyen CO_2 Lab project provides a unique starting point to study "greener" energy based on coal.

Climate issues were high on the agenda during the EU-US Summit in Brussels at the end of March 2014, when President Obama met with leaders from the European Union (Herman Van Rompuy, president of the European Council, and José Manuel Barroso, president of the European Commission). It comes as no surprise that they agreed that climate change is a risk factor for world security – a threat to sustainable economic growth. In addition, the parties agreed to exchange and strengthen expertise in the field of carbon capture and storage. Many acclaim this agreement and look forward to vigorous implementation of CCS in Europe, the United States and elsewhere around the world.

"Norway must cut faster. Emission cuts corresponding to 8 million tons remain if we are to attain the national goal for 2020. We will implement new measures and work continuously to find ways to reduce the emission of greenhouse gases," said Tine Sundtoft, Minister of Climate and Environment, in a speech after having received the 5th IPCC report in March 2014.

She went on to say, "I am confident that we can reduce our emissions and simultaneously adjust to a new future. A low-emission society is a good society to live in. A high-emission society is not an option." I approve of this statement – and I hope many others do too.

Let us hope that such statements will guide development both in Norway and internationally, and that Norway will contribute with good solutions as we approach the climate summit in Paris in 2015, so that we may reach binding agreements on how to reduce CO_2 emissions as much as possible.

Even as the Norwegian government is working on a new CCS strategy (a new full-scale capture facility) to reach the 2020 emissions goal, substantial funds are invested in European cooperative research efforts on capture, transport and storage of CO_2. That is great! CO_2 is a joint problem that requires cooperation across national borders if good solutions are to be achieved quickly. In this context, Longyearbyen CO_2 Lab may play an important role as a test site and knowledge base for CO_2 storage.

One thing is certain. If present emission levels are maintained for 20–30 more years without countermeasures, it will be impossible to attain the two-degree target by 2050. Coal is the main source of CO_2 emissions, and must therefore be part of the main solution.

"The solutions are many and allow for continued economic and human development. All we need is the will to change." These are the words of Rajendra Pachauri, IPCC chair and friend of Norway, after the most recent IPCC synthesis report was presented on November 1, 2014. Dr. Pachauri has visited Svalbard many times and shown great interest in the climate research and the work on CO_2 that is being done in the High North.

Hopefully, the coming years will see a sharper focus on the ultimate issues – the links between weather on earth and the resources available to humanity – and reach farther than far. Regardless of where we are, in the Arctic or at the Equator, we must have a care for life – fragile, precious life – as we race toward the future.

The first seeds were deposited in the Svalbard Global Seed Vault in 2008 (Photo: Mari Tefre / Svalbard Global Seed Vault)

The frozen ark

by Roland von Bothmer, Professor, Swedish University of Agricultural Sciences / Senior Advisor, Nordic Genetic Resource Center, and Ola Westengen, Coordinator of Operation and Management, Svalbard Global Seed Vault, Nordic Genetic Resource Center / Researcher, Center for Development and the Environment, University of Oslo

We Scandinavians naturally think about melting ice and rising sea levels when we consider climate change. As Lægdene and Helgesen write in their chapter "Ethics in the Arctic", the livelihood of Arctic peoples who sustain themselves by hunting comes under threat when the ice retreats. The concomitant sea level rise threatens low-lying coastal settlements all around the world. This relationship illustrates how the modern urban lifestyle impacts traditional communities – north and south. This is undoubtedly a major ethical dilemma, but as is so often the case with ethical dilemmas, we find it convenient to brush it aside because it "doesn't involve us". We cannot say the same thing about the threat to the climate that forms the backdrop for the project we are about to describe. It is becoming increasingly clear that one of the most serious consequences of climate change will be diminishing food production in the world. Given the Earth's growing population, it is beyond doubt that we face a huge challenge simply feeding everyone. To meet this challenge, we need agricultural research, and the most basic materials required for research and agriculture are what we call genetic resources – and this is what we preserve as frozen seeds in the Svalbard Global Seed Vault.

Adaptation to climate

In recent years we have heard much about adapting to climate change, both from climatologists and in the international climate debate. Some years back, many viewed the debate about climate adaptation as a defensive strategy intended to shift the focus from the most important task – reducing emissions. Now we know that there is already so much carbon dioxide in the atmosphere that a continued increase of global temperatures is inevitable. The world has become warmer, and will continue to grow warmer. This means that adaptation is no longer an alternative to emission cuts, but a necessary complement. The latest IPCC report about impacts, vulnerability and adaptation, released in March 2014[41], presents research from many fields related to adapting to climate change. The report shows the need for adaptation in many aspects of society. Infrastructure must be reinforced to tolerate more extreme weather; political and organizational measures must be taken to help those in greatest danger; research and development within enterprises that rely on natural resources must focus on adapting to the types of changes the climate models predict.

The part of the new report that attracted most attention in the media was the chapter on food security and food production systems.[42] That chapter described how food production has already declined in several parts of the world, a decrease that partially explains why global food prices have risen since 2007. The report mentions the probability of a decrease in total agricultural yield of up to 2% per decade from 2030 on, unless farming is adjusted to the new conditions. This negative trend will coincide with a 14% increase per decade in the total demand for food, up until 2050. The predicted effects on the productivity of various crops in different regions of the world depend on which emission scenarios and climate models are used, but it is clear that any potential increase of productivity in temperate and arctic regions will be entirely inadequate to compensate for loss of productivity in the tropics. The most severe effects will be seen in areas that are already marginal for farming because of drought and extreme temperatures. The predicted decrease in production of maize in southern Africa and rice in India are dramatic examples, and will have negative impact on food security in these regions.

But even though the prophecies are gloomy, the report emphasizes that negative effects can be averted, provided the measures taken to adapt are successful. Many such measures must be taken at the political and institutional level: examples include establishing irrigation systems and improving weather forecasting. Other measures can be taken by the farmers themselves, such as adjusting the time when they sow their seeds and switching to hardier crops. According to the IPCC report, it is this last climate adaptation measure – switching to crops with better tolerance for the new climate conditions – that will have the greatest positive effect. Ultimately, agriculture cannot adapt to climate change until the crops themselves have been adapted.

There is something boundlessly unfair in the fact that poor countries in the south will suffer the greatest food insecurity owing to climate change. Perhaps it is this boundlessness that inspired the world community to adopt a resolution at the UN General Assembly in 2009, stating that it is a global duty to develop crops with greater tolerance for drought, flooding and other stress factors related to climate change.[43] Within international foreign policy, it is no novel insight that development of new crops is an important means of ensuring social advances. In 1970, the American plant breeder Norman Borlaug was awarded the Nobel Peace Prize for contributing toward world peace by helping increase global food production. Norman Borlaug was one of the people who developed the new crops that swept through Latin America and Asia in the "Green Revolution" of the 1960s. It is now often claimed that the green revolution was so successful at increasing food production – and thus reducing food prices – that the world "forgot" to provide public funding for agricultural research and development in northern and southern climes between the 1980s and a few years into the 2000s. But since what has been called the food crisis began in 2007, agriculture and food security have once again climbed to the top of the political foreign aid agenda. It was during this new era, when the importance of agricultural development was acknowledged and the field's status grew, that the Svalbard Global Seed Vault shouldered its role as a backup storeroom for the world's genetic raw materials – the diversity of genes and combinations of genes contained in seeds from around the globe.

Genetic diversity is one component of overall biological variability, often called *biodiversity*. Biodiversity is often described as having three levels: ecosystems, species, and genes. While most of us are now aware that pivotal ecosystems such as the rainforest, and charismatic species like the polar bear are threatened by human activities, fewer realize that the genetic diversity in common crops is also threatened. All three levels interact. When an ecosystem is threatened, the species that live in it are threatened. When a species is endangered, its genetic diversity declines, and when the genetic diversity within a species is reduced, the species itself becomes vulnerable. For if there are any laws of nature in biology, one of them stipulates that genetic variation is a prerequisite for evolution – nature's continual adaptation. The raw material for continued evolution of plants – even for ubiquitous, mass-produced crops like wheat, rice and maize – is the genetic diversity found in older cultivars and their wild botanical cousins. This role as the raw material for continued evolution has given rise to the technical term for genetic diversity: genetic resources.

International efforts to preserve genetic resources took off in earnest in the 1960s. In the wake of the success of the super-productive crops of the green revolution, influential researchers warned that the genetic diversity of the traditional crops was in danger of being out-competed and lost forever. The result was genetic erosion – inadvertent loss of genes in the cultivars that were no longer grown when the modern varieties took over. There are many reasons – both ethical and economic – why humankind should try to prevent loss of biodiversity. The reason we must conserve the genetic diversity of our crops is existential: without this diversity we will soon stand without our daily bread.

Plant breeding: controlled evolution

To explain the role the Svalbard Global Seed Vault plays in this era of human-induced climate change, we will first briefly describe the miraculous link between genetic resources and food production: plant breeding. It is a paradox that plant breeding is something most people know nothing about (even though its results lie on their dinner table), that it is chronically underfinanced, and that many view it with skepticism – as a way of "messing with nature". In part, this is probably related to recent developments in plant breeding, but much can also be attributed to the fact that we now take it for granted that our crops will give a bounteous, dependable harvest.

Simply put, plant breeding involves selecting plants with desirable traits, over several generations. If the trait being selected for is inheritable, new generations will change in a desirable direction. The first farmers were also the first plant breeders. One of the traits of wild plants that the first farmers came to grips with was spontaneous seed shattering. For wild grasses, it is most advantageous if the ears all ripen at different times and the seeds are blown off by the wind. For barley – which is a domesticated grass – it is much more practical if all the ears ripen simultaneously and the grains stay attached to the ears until the farmer has harvested the crop with a sickle or combine.

Over the thousands of years that have passed since the first plants were domesticated, our crops have continued to evolve, and many of them now bear little or no resemblence to their wild relatives. This genetic change has come about both through natural selection and adaptation to climatological and ecological conditions, and through selection guided by human preferences. People have chosen the varieties that tasted better, lasted longer in storage, produced more, or grew better. This led to development of a multitude of locally optimized varieties, sometimes called landraces, of all our important food crops. When scientific plant breeding began in Europe in the 19th century, the landraces served as the starting point. They were genetically diverse and well adapted to local conditions. Later on, plant breeding relied on crossing different lines or varieties with each other to create new, valuable combinations of traits. This is why genes from traditional landraces survive in modern crops.

Over the last hundred years we have seen major advances in genetic science, which have also affected plant breeding. One of the most important breakthroughs came with the development of mathematical and statistical methods to design experiments and guide subsequent selection. Later a number of biotechnological breakthroughs made it possible to manipulate the natural variability to generate new characteristics. In mutation-driven breeding, researchers used radioactivity or chemicals (mutagens) to generate new mutations in the genes. Cell and tissue culture involved manipulating single cells (protoplasts) and regenerating entire plants from them. The molecular biological revolution that began in the 1970s and is still going on, also led to many methodological breakthroughs. By using molecular markers, scientists can easily select plants that possess the genes of interest without being required to wait until the plants resulting from a cross have matured. It is also molecular biology that has made it possible to transfer genetic material from one species to another, creating transgenic plants, also called genetically modified organisms (GMO). Genetically modified plants currently on the world market have been given genes that increase their content of certain nutrients, or protect them against diseases or pests. This protective effect can be direct or indirect, such as making the plants less sensitive to herbicides, so that only weeds are killed. GMO is a constant topic of debate; European authorities strictly regulate use of transgenic plants, whereas countries such as Brazil, Argentina, India, and China have more liberal policies, along with the United States, which is the country where this technology is most commonly used in agriculture.

In the past hundred years, there have also been major changes in how plant breeding is organized and who owns the seed companies. Initially, plant breeding was mainly done by small family firms or government institutes and universities. When farming became more profitable, a lucrative seed market developed in some countries, attracting large private enterprises. In countries like the United States, small family-run operations and the most lucrative biotechnology firms have been bought up by huge multinational corporations. These giant corporations mainly breed and sell seeds used in large-scale commercial farming, crops like maize, soybeans and cotton. In less lucrative markets the multinational corporations are not equally dominant. There is still some publically financed plant breeding, and small seed companies continue to develop varieties for places with special farming needs, such as the Scandinavian countries. For the countries we initially described as most likely to suffer loss of productivity owing to climate change, plant breeding is still predominantly done by publically funded institutes. The international research centers that were central players in the green revolution are mainly financed through foreign development aid, and they provide much of the raw material that is later adapted to local conditions by national agricultural research institutes and local seed companies.

The steady development of plant breeding in the past hundred years does not mean that old techniques have been left by the wayside. Combination-breeding (by crosses) and selection still lie at the core of all plant improvement programs, though they are now supplemented with other techniques as required. Plant breeding, despite all the new methods, is a slow process and it still takes eight to ten years to develop a new variety – about the same amount of time as it took a hundred years ago. New developments have likewise not altered the fact that the raw material for all plant breeding is the genetic diversity found in the traditional varieties and their wild relatives – regardless of whether breeding is done the old fashioned way in the farmer's field, or combined with modern scientific methods.

Genetically diverse landrace of barley in the Gilgit province of Pakistan near the Chinese border (Photo: Roland von Bothmer)

Gene banks: frozen genes

Gene banks, like botanical gardens, are places where species are preserved *ex situ* – outside their natural environment. The history of gene banks begins with the Russian geneticist Nikolai Vavilov, who realized around 1910 how valuable and vital it was to preserve and make use of genetic variation in plant breeding. Over a period of about 25 years he and his co-workers traveled around the world, gathering whatever they could find in the way of crops and their closest wild relatives and shipping them home to Russia. This collection was the beginning of what is now the world-famous gene bank in St. Petersburg, which bears Vavilov's name. Although similar collecting ventures were undertaken by researchers in other countries, particularly in the United States, it was not until the 1960s that the concept of establishing gene banks took hold internationally. At that time, intensive collection efforts began worldwide and many national and international gene banks were established.

One of Vavilov's hypotheses was that the world's main crops had been domesticated in a limited number of regions, and that these centers of origin could be identified by seeking out the areas with greatest variation. Vavilov identified eight primary centers of diversity, and his map was later refined by other researchers studying cultivated plants. The temperate zone crops that are most vital for us in Scandinavia emanate mainly from an area in Southeast Asia called the *Fertile Crescent*. Many important crops such as wheat, barley, peas, lentils, faba beans and flax were domesticated in this region. This is where farming began, where humans first settled down and laid the foundation for the world's first civilizations. When these plants spread to other parts of the globe, they gradually adapted to new environments, and new combinations of traits developed. In this way, several secondary centers of diversity arose, for example centers for wheat and barley in Ethiopia and central Asia (Kashmir and Tibet). To this day, the widest and perhaps the most important genetic variation is found in these primary and secondary centers of diversity. Two features common to several of these areas are that they are inaccessible and politically instable.

Ever since collection of genetic resources began in earnest in the 1960s, many national and regional gene banks have shouldered the responsibility of collecting and preserving seeds from their respective territories. In the Nordic region, the five countries of Denmark, Finland, Iceland, Norway and Sweden, working together within the Nordic Council of Ministers, established the Nordic Gene Bank in 1979. Today the institution is called the Nordic Genetic Resource Center, NordGen, and works with farm animals and forests in addition to running the Nordic Seed Bank in Alnarp, Sweden. The major international agricultural research institutes that were central players in the green revolution, share the responsibility of preserving the plants they do research on, and which they try to improve. The most renowned institutes are the International Maize and Wheat Improvement Center in Mexico (CIMMYT), the International Center for Agricultural Research in Dry Areas (ICARDA), and the International Rice Research Institute (IRRI) in the Philippines.

According to the Food and Agriculture Organization of the United Nations (FAO), the world currently has 1750 gene banks, large and small. These gene banks report a total store of over 7.2 million seed samples. The technique required to preserve seeds is fairly simple: the seeds must be dehydrated to low moisture content, sealed in a container (usually a bag made of plastic-coated aluminium) and stored at –18°C. But even though an ordinary freezer serves fairly well for seed storage, several other aspects must be taken care of in a responsibly managed gene bank. The seeds' ability to germinate declines with time, so the gene bank must test this regularly; must sow and cultivate the old seeds and generate new ones when those in the original sample are no longer able to germinate satisfactorily. In addition, preservation is not the only objective of gene banks: they also offer a service to users, who are mainly plant breeders and other researchers, but also some farmers and gardeners. It is therefore essential to maintain as much relevant information as possible about what is being stored. This means that the gene bank must save data about how, where, and when the seeds were collected, and generate new data through field studies aiming to characterize and evaluate the seed sample. Access to such information enhances the demand for seed samples, and well-run seed banks offer users a searchable database and a simple procedure for placing orders for seeds. Last but not least, a gene bank must ensure that there is a backup copy of each sample in a different geographical location than the original sample. This is the role played by the Svalbard Global Seed Vault: it is a backup storage facility for the world's gene banks.

It is obviously a good idea to have a backup. Many of the world's gene banks are situated in dangerous places, prone to natural and man-made disasters. There are many examples of gene banks being devastated; for example, the gene bank in the Philippines was laid in ruins by a typhoon in 2006, and the Egyptian gene bank was looted in the wake of the 2011 revolution. Some of the most precious seed collections are stored near the areas Vavilov and others identified as centers of diversity. Since many gene banks lie in developing countries, they are threatened not only by dramatic events such as natural disasters and war, but also by ordinary everyday problems like power outages and sudden withdrawal of funding for operations. Consequently, gene resources that are endangered or already extinct in the wild and on the farm are also vulnerable in many of the world's gene banks.

Svalbard's Ark

The notion of establishing an international backup collection in the permafrost of Svalbard dates back to the 1980s, but it would be twenty years before idea became reality. The reason it took so long is that the topic of genetic resources is strongly political; it has been and still remains a cause of international contention and debate. The question of who owns the seeds deposited in the world's seed banks became an inflamed topic following the privatization of plant breeding we described earlier. To put it succinctly, private seed companies want to have "copyright" to the varieties they develop, and the countries that provided the genetic raw materials for plant breeding want "royalties". Ten years of negotiations under FAO eventually resulted in the International Treaty on Plant Genetic Resources for Food and Agriculture, which is now the most important international agreement regulating preservation of and access to genetic resources, and distribution of any benefits generated using these resources as a starting point. At the core of this consensus lies the realization that all countries are mutually dependent where access to genetic resources is concerned. The diet of most of the world now includes plants that mainly originated far away. Consider, for example, that maize from Central America is now the most important grain in Africa, and that essentially all the food crops in Scandinavia have been "imported" at some time in history. Continued improvement of these plants relies on access to genetic resources from other countries. The "Plant Treaty" also strengthened confidence in – and willingness to cooperate in – seed storage itself, and it became possible to establish the seed vault that many had long envisioned.

An international commission was established, headed by Cary Fowler, who was then professor at the Agricultural University of Norway at Ås. Fowler concluded that Svalbard was an admirably suitable site. The most important reasons the commission listed were: 1) Svalbard is in a peaceful part of the world, and the Svalbard Treaty prohibits military activities in the archipelago; 2) Norwegian authorities have the confidence of international political players and are considered sufficiently neutral in the international debate about ownership of genetic resources; 3) The local community of Longyearbyen is run exceptionally well, and local authorities are highly competent; 4) The permafrost ensures low temperatures so that even if the refrigeration unit breaks down, the seeds would never be exposed to detrimentally warm conditions; 5) Svalbard is fairly stable geologically, and placing the vault deep in the rock provides extra security.

In the summer of 2006, the Prime Ministers from the five Nordic countries participated in an event marking the Norwegian decision to construct the seed vault, and in February 2008 the vault was opened in the presence of guests from near and far. Norwegian Prime Minister Jens Stoltenberg, Nobel laureate Wangari Maathai, European Commission President José Manuel Barroso, and FAO Director General Jacques Diouf were among the many prominent guests who helped carry the first boxes of seeds into the vault. Even before it was inaugurated, the seed vault had attracted massive attention in the media, not least because Cary Fowler (who by that time had become director of the Global Crop Diversity Trust, which was the international partner in the project) had done excellent outreach work in this early phase. Some claimed that the media presence in Longyearbyen at the time of the opening was stronger than it had ever been since Roald Amundsen disappeared in the ice while searching for Umberto Nobile.

Today the seed vault is a shining landmark on the mountainside above the airport and can be seen by anyone who arrives in Longyearbyen in the dark season. The characteristic gateway leads down a tunnel that opens onto the transverse passageway that gives access to the three storage vaults 120 meters deep in the mountain. The Longyearbyen branch of Statsbygg (the organization that runs public sector properties in Norway) maintains the facilities and is in charge of security, whereas NordGen is responsible for all the practical and administrative issues involved in the shipping and storage of seeds. Operating costs are covered by Norway through the Ministry of Agriculture and Food, and the Global Crop Diversity Trust.

Most of the year, the seed vault lies dormant. But three or four times per year we open the many steel doors to deposit new boxes of seeds that have been packed and shipped from the gene banks that use the seed vault for backup storage. In our contract with the gene banks, we make it clear that the seed vault functions in the same way as a bank box: the seeds will never be given to anyone except those who sent them in the first place.

The numbers show clearly that many countries and gene banks now trust the seed vault's ability to preserve the seed samples in safety. At the time of the vault's fifth anniversary in 2013, we estimated that about one third of the precious samples preserved in the world's gene banks were already duplicated for safekeeping in the seed vault.[44] The largest collections have been sent from the international network of gene banks, but a number of national and local gene banks have also deposited seeds. There is something magnificent about seeing boxes of seeds from countries as different Burkina Faso, Mongolia, the United States, Germany, and North and South Korea stored side by side in a single room. Even if the cooperation is indirect, this collection sends a strong signal that our joint world heritage is preserved deep in Platåfjellet. During his visit to the seed vault in 2009, UN Secretary General Ban Ki-moon said:

> This is a very creative initiative to fight for food security in the face of climate change in a much longer term. The seeds stored here in Svalbard will help us do just that. They come from virtually every country in the world. They contain the essential characteristics that plant breeders and farmers will need to ensure that crops become climate-ready and even more productive. (…) Sustainable food production may not begin in this cold arctic environment, but it does begin by conserving crop diversity.

This quote shows that the importance of preserving and making use of genetic resources is acknowledged at the highest political level, and we believe that the seed vault has played a crucial role in putting this topic on the agenda. One might say that the seed vault plays two roles: it is a backup storage facility for gene banks around the world, and it is a symbol for the entire issue of genetic resources.

These roles have synergy effects, and the seed bank's international prestige can be used to move the project forward. Several countries that have been hesitant to send genetic resources outside their own territory have now chosen to deposit seeds in the safe haven in

Svalbard. Along with other partners in the project, we are working hard on what might be called seed diplomacy – instilling confidence in international collaboration to preserve genetic resources. A smoothly functioning system for preservation and use of genetic resources is a prerequisite for food security in the world. Recent events in Syria have given us a dramatic reminder of why the Svalbard Global Seed Vault is so vital. As mentioned earlier, many of our most important crops originate from the region that is now Syria. Much of the diversity – both wild relatives and old landraces – has been gathered and stored in the ICARDA's international gene bank on the outskirts of Aleppo. Today this gene bank lies in the midst of a war zone. The personnel who remain on duty are making heroic efforts to preserve the seed collection. An appreciable proportion of the collection had already been deposited for safe storage in the seed vault before war broke out, and over the past year, staff members have managed to send another thousand seed samples to safety. These seeds are priceless natural and cultural heritage, shaped over millennia in the region that is often called the cradle of civilization. We will need this heritage for adapting to new times.

Living conditions are challenging for arctic plants and animals, and the ecosystems are vulnerable to climate change (Photo: Venke Ivarrud)

Vulnerable life

by Leif Magne Helgesen, Pastor of Svalbard

Snowflakes fall silently. We are in what looks like a romantic Christmas card. A few people are out skiing in the polar night. The lack of sun and light in these winter months creates a long, dark night. It isn't always as bitterly cold as one might expect this far north, but the wind can easily pick up and send the windchill factor plunging.

It would be impossible to establish a community like Longyearbyen in other parts of the High North. This close to the pole there is usually ice, but the Gulf Stream outside the west coast of Spitsbergen gives a warmer climate than at similar latitudes in northern Canada or Russia.

Warm currents from the south influence the amount of ice on the fjord. It is no longer commonplace to see ice on Spitsbergen's largest fjord, Isfjorden. This fjord is the entryway into Longyearbyen. Before Longyearbyen airport was completed in 1974, making it possible to fly up in the winter, the fjord was the main thoroughfare. "The last boat" used to be an established concept; it meant that the townspeople would be isolated until the ice on the fjord broke up, allowing boats to approach and re-establish contact with the outside world. In recent years, Isfjorden has been ice-free in winter. The abnormal has become normal.

Ice-free fjords do not necessarily result from climate change, as ocean currents can influence local ice conditions. Year-to-year variations need not be related to global climate change. And we mustn't let a single measurement of snow depth serve as a global thermometer. If it doesn't snow in Longyearbyen in September, that is not the end of the world. Lack of ice on the fjord doesn't prove that the sand is running out in the hourglass of humanity's time on earth. Doomsday prophecies often turn out to be premature, regardless of whether they are made by religious groups, scientists, or the press.

Similarly, we cannot dismiss the climate crisis even if ice should form and those of us who live here could climb onto our snowmobiles and take a shortcut across the fjord. In the same way, we cannot dismiss the climate crisis because snow conditions are good and the cold nips our faces. Climate change must be measured over time and distance. Nonetheless, we who live in Svalbard discern a negative local trend toward less ice and more precipitation. These are signals from nature and we must pay attention.

It is easy to trivialize the climate situation. Life is simpler if we pretend the outside world is irrelevant. Relating to the future is just as tricky as relating to millions of poor people who are already feeling the impact of climate change. Effects that climate change has on other places, but that leave us untouched, are easy to ignore. It's most expedient to feign ignorance and not bother.

Fear of the future is not the overriding emotion most of us carry in our hearts. We are more prone to indifference and lack of willingness to chip in for the sake of generations to come. Most of us think only of today – and even that can be a struggle. If forced, we may manage to plan a year into our own future. Looking far ahead to discern the consequences our own lifestyle on future generations is a far more difficult task. It takes courage, willpower, and an ability to take responsibility.

Even politicians avoid thinking farther ahead than their own terms of office. It is more important to be re-elected than to take responsibility for what might happen years into the future. It is more important to attain growth and prosperity in the constituency than to invest time and effort in what is already happening in other places. This is especially clear in American politics, where the president can be more daring in the second and last term, when he knows he cannot be re-elected. The threat of not being re-elected often impedes urgent, necessary changes – particularly when global issues are at stake.

The church as an institution embodies both the possibility and the capability of looking far into the future. Religious leaders are not elected for four-year terms; they are free to think ahead. Their perspective stretches to eternity. Ethics requires far-sighted thinking. We owe something to our children's grandchildren. Ethics challenges us to look beyond the horizon. With current knowledge as a foundation, we are obliged to act today if it can save lives fifty years hence.

How well the world's churches and their leaders acknowledge responsibility toward the climate varies. One risk is too strong a focus on life after death. A church that doesn't take life *before* death seriously is neglecting its responsibility and its calling. As a pastor, I hope

Leif Magne Helgesen, Pastor of Svalbard, conducting worship service on Hiorthfjellet (Photo: Torbjørn Gilberg)

for eternity, and I believe in life. I ascribe to what the British humanitarian organization *Christian Aid* says in its slogan: "We believe in life before death".

Church bells call from the Arctic

At the time of the climate summit in Copenhagen in 2009, a church bell relay started from the church in Longyearbyen, Svalbard kirke. The idea was that church bells all over Norway would ring twelve strokes for the climate summit. In Longyearbyen there was a torchlight parade and a ceremony with children's choir, appeals, prayers, and songs. The church and the research institutes stood united in this ceremony.

Twelve strokes chimed from the church bell as a symbol of danger, but also as a sign of hope for change. We must cherish hope as a means to reinforce optimism and energy. If the picture is painted too black, no one will see the contours of what is possible. An endless, lightless tunnel would soon lead to despondency and a feeling that effort is to no avail.

Twelve strokes symbolized that this is the twelfth hour: that changes and international agreements are required. A call went out from the Arctic to the delegates at the climate summit, with prayers that they would stand together for the environment. The call went out on behalf of people all over the world who will suffer, each in their own way, unless the world's leaders can unite around actions and agreements.

The world's northernmost newspaper *Svalbardposten* has often trivialized the threat to the climate and looked askance at research and international forums. The week after the church bell relay began, the editor wrote:

> We rejoice for every degree below the freezing point, study the latest ice charts and note that there is already solid ice at the inner end of the fjords. Oh, yes, we sense that this year, conditions will be fine. Don't shout at us about a global hell, eternal torment, and a hundred strokes of the cane. We sense it in the north wind. Soon the cold will arrive in sharp gusts. We tie our fur hats firmly on and get ourselves ready. (*Svalbardposten*, No. 48, 2009)

We would all be delighted if the climate crisis could be canceled. We would be happy if thousands of scientists around the world were mistaken, and the earth's temperature regained its balance. The fact that there have been errors in the scientists' analyses and information gives food for thought and weakens trust in their climate analyses and reports. Nonetheless, it doesn't eliminate the seriousness of the situation, or the consequences that many people are already experiencing in the form of more frequent floods, more drought, and more extreme weather.

Unfortunately, we can't cancel the crisis, even when ice forms in the inner fjords. The climate issue is no less serious simply because an editor pokes fun at it. Nor will it go away because no international agreements are reached, or superpowers such as the United States and China choose not to enter into international agreements. Their cowardice has to do with concern for their own economy and political future rather than concern for responsibilities and the world's future.

Politicians are accountable to their constituents and their compatriots. Editors, religious leaders, and business executives are also accountable. The question is who dares to think big, and who will shoulder responsibility for the world's citizens and future generations.

Petition read during the ceremony beside the church bell:

Call from the Arctic

Sea ice in the Arctic Ocean is melting. The glaciers in Svalbard are retreating. Sea temperatures are rising along with atmospheric CO_2 levels. This concerns us all.

The earth's climate is an integrated system based on energy from the sun. Regions near the equator receive most energy and polar regions receive least. The climate depends on the ability of the oceans and the atmosphere to transport energy from the equator to the poles.

When we change the composition of the atmosphere, we also change basic conditions for how the atmosphere functions in the climate system. Increased levels of greenhouse gases lead to warming in oceans and on land.

The Arctic is a barometer for the climate all over the earth.

The Arctic is particularly sensitive to warming. Large areas are covered with ice and snow, but the temperature is near the melting point. Small increases can cause dramatic alterations. Changes in the Arctic will lead to changes in the rest of the earth's climate system.

Agriculture relies on a predictable climate. Different societies grow different crops, depending on the local climate. The poorest among us are also those most vulnerable to changes.

This means the solutions must reflect global justice within a dimension that has so far not been incorporated into international agreements.

The changes have an ethical dimension that touches all of us.

In the Arctic, we can already see that living conditions are changing. Plants and animals struggle to survive. Indigenous peoples can no longer hunt as they used to. This changes the very basis for their existence.

We send a call from the Arctic to the delegates at the climate summit, with a prayer that they will stand together for the environment. We do this on behalf of people all around the world who will suffer, each in their own way, if the world's leaders cannot agree on measures that will make a difference.

Before it is too late!

Longyearbyen, November 2009

Signed by
Leif Magne Helgesen, Pastor of Svalbard
Gunnar Sand, Managing Director, University Centre in Svalbard
Kim Holmén, Research Director, Norwegian Polar Institute
Stig Lægdene, Principal, Northern Norway Educational Centre of Practical Theology

The precautionary principle

Svalbard is a world heritage managed by Norway. We who live here treasure all the glory of nature. We enjoy ice and sub-zero temperatures. When the weather forecast shows increasingly frigid blue numbers, our pleasure and well-being also increase. That is as it should be. We aren't upset if the same forecast shows summer and sunshine in the south. The world needs a cold climate up north.

Every degree below freezing is a good thing. If someone is picking strawberries in mainland Norway while a snowstorm sweeps through Longyearbyen, that's a sign that the world is in some kind of balance. It would be worse if trees and cloudberries started growing in Svalbard. Luckily, it will be a long time before we can chop our own Christmas trees in Longyearbyen, even though influx of new plant species to the Arctic is one possible development when the climate changes.

Svalbard's summer is brief and hectic. Picking flowers and plants is prohibited. That says something about the vulnerability of the flowers that emerge when the snow melts in June. Each plant has a brief time span before the next snow flurry paints the landscape white again. Like satellite dishes they absorb as much light and heat as they can in the short growing season.

The plants speak volumes about life's vulnerability, but also about resilience. The plants are vulnerable and hardy at the same time. When the first saxifrages peep out between mounds of snow, they demonstrate the power of life amid the barren landscape.

The plants convey to humans something about being simultaneously vulnerable and resilient. Like the plants, we are vulnerable, but we are not weak. We can change the future. We can join in fellowship and make life better for people living today and in centuries to come. That shows strength and gives perspective.

When climate changes, it touches our entire existence. We know that living conditions have changed before, but humanity has never before influenced our surroundings the way we do today. That gives us both responsibilities and opportunities.

The protected status accorded to all plants in Svalbard reflects the strict environmental regulations the Norwegian government enforces through the Governor. The Svalbard Environmental Protection Act is one of the strictest in the world. The objective of the strict management in Svalbard is described thus:

> Norway's overarching objective for environmental protection in Svalbard is to preserve the archipelago's unique wilderness. Svalbard shall be one of the world's best managed wilderness areas. Both the Svalbard Environmental Protection Act and related regulations shall protect the vulnerable environment and regulate activities in Svalbard. (www.sysselmannen.no)

The Svalbard Environmental Protection Act is a central management tool in the archipelago. An important principle lies at the core of this Act: the precautionary principle. This entails that if knowledge about the consequences of a certain activity is inadequate, protection of nature has top priority. If human activities are detrimental to the environment, the activities much be restricted or stopped altogether.

Whether this principle will extend to the seas around Svalbard is more doubtful. Principles are often watered down in proportion to availability of natural resources. Here the "dollar principle" has supremacy. Many nations are keeping a close watch on developments in the north. Interest in these high latitudes grows with each new asset that is discovered. Above all, the possibility of hidden reservoirs of oil and gas increases interest in the Arctic. Several nations and large corporations are jockeying for a position in the High North.

"In the long term, we will have to move petroleum operations successively farther north. We must start working on the possibility of discovering oil and gas in Barents Sea North, around Svalbard, and around Jan Mayen," said Per Terje Vold, Managing Director of the Norwegian Oil Industry Association, in the report "Oil and gas activities in the north" from 2009. These words reflect an offensive strategy that continued in the years after the statement was made.

Prospecting and exploitation of natural resources around Svalbard raise many questions about environmental safety. It is fine if the earth's resources can be exploited in a responsible way, but do we have the technology required to prevent major spills in this exquisitely vulnerable region? The major stress caused by extreme weather and ice conditions requires new technology to prevent environmental catastrophes. The question is how much we want to preserve for the future, and how much we want to exploit for short-term profit.

Through the centuries, Svalbard's natural resources have frequently been exploited ruthlessly. Hunting brought many species to the brink of extinction. Today all hunting is strictly regulated by environmental statutes. Walruses and polar bears are protected, and whale hunting is restricted to a few areas and species. As a result of protection and quotas, populations are growing again.

But now perhaps history's pendulum is swinging back. At times, polar bears, whales and walruses were slaughtered. The populations were threatened by people in search of money and short-sighted gratification. Today our exploitation and use of fossil fuel is a major factor threatening the survival of some species. Once again, life is endangered by people's search for short-term wealth and happiness.

Extraction of oil and gas in Barents Sea North would provide many new jobs and contribute significantly to the world's energy supply. Longyearbyen could potentially grow if the town becomes a central operational harbor and a hub for all the infrastructure required. Then again, extraction could cause irreparable damage and devastate one of the world's best preserved regions. The inheritance we are expected to pass on would be ruined. The greatest danger lies in perturbing the natural balance and raising temperatures even more.

This brings us back to the issue of international regulations and agreements. Vast riches in the form of petroleum can get in the way of ethical reflection. Dollars have never been the best indicator that enduring values will be given optimal conditions for survival. It would be worrisome if short-term economic prosperity trumps long-term prudent management of natural resources in the Arctic.

Other ethical dilemmas follow in the wake of growing interest in Svalbard and the adjacent seas. Every human activity in the High North should be preceded by thorough reflection on ethical issues. It is also vital to maintain a far-sighted perspective. This is about the future of the earth and of humanity.

Coal

Coal mining has been the economic mainstay and main rationale for communities in Svalbard since the beginning of the 20th century. Longyearbyen was founded by John Munroe Longyear in 1906. Since then, prospecting and mining coal has been the main industry in Longyearbyen; the same has been true of the Russian communities on Spitsbergen.

In 2013, the mining company Store Norske Spitsbergen Kulkompani opened a new mine in Lunckefjell. This will mean a lot to the local community in Longyearbyen, at least for a while. The fact that the already established mining district in Svea will be used for logistics is crucial, as it precludes the necessity of building new infrastructure. Mining operations in Svalbard are nonetheless an ethical challenge, posing the dilemma of leaving permanent tracks in a vulnerable area. We must always ask ourselves whether the detrimental effects of mining will be too great and long-lasting. At the same time, all the coal in Svalbard would not contribute significantly toward covering global energy needs. The deposits and the mining operations are too small.

Mining in Svalbard has a limited future. A coal mine is only productive for a short time. The major deposits appear already to have been dug out. The question is both whether mining has an economically viable future, and if it is ethically defensible. Ultimately, the politicians must either make the decisions required for continued coal mining, or stop the mining altogether.

Svalbard's coal is renowned for its high energy content, and is considered "clean". Nonetheless, coal mining is controversial. The industry must and should adhere to strict environmental regulations both locally and globally.

It is only a matter of time until coal mining in Svalbard is history. Then research, tourism, and a potential operations harbor for shipping in the Arctic will probably be the mainstays of continued activity and development of the communities on Spitsbergen.

Mining in vulnerable areas is controversial. Black coal in the beautiful white landscape can evoke strong emotions. Coal mining in Svalbard must also be viewed in the context of political presence. For over a century, retrieval of the black gold has created and maintained a Norwegian population in the archipelago. Despite this, coal mining is probably going into its final phase. The deposits – and thus also the profits they generate – are too small. At the same time, coal mining is incompatible with Norway's goal to reduce CO_2 emissions. Just as hunting now belongs to Svalbard's history, coal operations in Svalbard are likely to go into their final phase and become history. It is up to the mining company Store Norske and local business enterprises to redefine themselves so that the community in Longyearbyen can evolve. Anything else would mean walking backwards into the future.

The consequences of coal operations are nonetheless smaller and more local than those of potential petroleum operations around Svalbard. An oil spill in the waters off Svalbard would cause far more devastation than a few lumps of coal from the depths of the mountains around Svea and Longyearbyen. Primary production in the ocean takes place at the ice edge. This means the region teems with life both in the water and the air, and in the ice itself. It would take very little to perturb that vulnerable life.

In the world's most beautiful cathedral: a makeshift altar on Hiorthfjellet overlooking Isfjorden (Photo: Leif Magne Helgesen)

The precautionary principle should apply also in the areas around Svalbard. Science and technology give opportunities. Jobs are important. Businesses must have sustainable conditions. The natural resources have been entrusted to us to manage responsibly. At the same time, we must think ahead, taking precautions when we are less than one hundred percent certain what the consequences might be. The possibility of protecting important pristine environments is more valuable than using up the resources of the High North over a short time span. We need to use the world's bounty in a responsible way.

The church in the Arctic

It isn't easy being a prophet. Even for men and women of the cloth it is a risky undertaking. And for scientists it is impossible. All research results are fraught with uncertainty.

No one knows the future, even if we have prognoses and reports showing trends and possible developments. Both human-induced and natural variations must be taken seriously. In this context we must let science take the lead in the search for answers.

It is imperative to take the warning signs seriously. We must take our future seriously. When we shirk responsibility and neglect to take necessary steps, we are being egoistic. The greatest threat lies in indifference and inaction. Our ethical choices should be rooted in a sense of community – community in a global perspective, stretching beyond our own time to generations to come.

As the world's northernmost church, Svalbard kirke is encircled by vulnerable arctic nature. That gives us an obligation. We are the church where the ice is melting. This is about life in Svalbard and the impact of climate change on animals, marine life forms, plants, birds – and on humanity. We all depend on each other in an unbreakable chain.

During services in Svalbard kirke, one verse in our confession reads:

> All the scorched fields on earth. Each tree that no longer bears leaf. Each bird that lies poisoned on the ground. Each fish that no longer swims in clean water. The ice that is melting. God, the fault is ours alone.

For the church, prayer is an action. It sets a framework for our deeds. Prayer puts into words the values the church wishes to live by. Prayer must not be a pretext for inaction, but a clarion call: a call to action. In Svalbard kirke, the intercession includes a prayer about life's vulnerability:

> We pray for all living things threatened by extreme weather, melting ice and rising seas. We pray for indigenous peoples and others put at risk by climate change.
>
> We pray especially for the rivalry in the High North. O God, you see humanity's hunger for more oil, gas, coal, fish and other resources in the north. Teach us to make use of the earth's bounty without squandering our children's inheritance. Give the world's leaders the wisdom and courage to make agreements that will benefit everyone on earth, now and for generations to come.
>
> God of love, who is always on the side of the meek, we pray for the world's poor, for those who will bear the heaviest burden of the climate change that we in the wealthy parts of the globe are mainly responsible for. Teach us to share our wealth. Teach us to strive for justice and peace. We pray for hope and a future for all human beings.

The church has an obligation. We must protest when life is under threat. It is our ethical duty. This is why the climate is a central concern for Svalbard kirke and churches around the world.

It is essential that the Christian church cooperates with other religious communities. Climate issues concern us all, regardless of faith. Acting on the climate is not a task for individual countries and experts. When temperatures rise and weather changes, it will be felt across all national borders, by all political factions.

The church and other religious communities have a perspective that stretches onward, ever onward. Our frame of reference runs from ancient times, through the present day, into the future. In the context of climate, this is important. Religious communities have a particular duty to speak up on behalf of life when others remain silent, a duty to support the forces of good that cherish life. We should not fear each other, but work together across the boundaries of nation and faith.

Clearly, religious communities and churches should also cooperate with researchers and the natural sciences. These groups have often viewed each other with distrust and kept a distance. We see similar distrust among some politicians and business interests.

Some groups believe they can meet the challenges alone, ignoring everyone else. That is a dead-end road. The climate is so great a challenge that humanity cannot afford *not* to work together.

The climate issue is the greatest ethical dilemma of our time, along with the issue of poverty. The climate threat is so dire that it requires courageous decisions and actions. It challenges every one of us: individuals, churches, businesses, politicians, and the entire community. The forces of good should unite to put in place binding agreements that will reduce emissions and focus on technologies that will protect the environment.

We must take the threat to the climate seriously. Focusing entirely on money and material gain in the current situation is immoral. We need to understand that solidarity is part of our duty toward the planet.

The alarming situation we now see raises questions about our lifestyle. We must be willing to take the steps necessary to prevent further deterioration of the environment. It is crucial for us all to protect vulnerable life. By doing that, we will create hope for the future.

One of the main dividing lines in politics goes between growth and conservation. It is alarming to see how economic growth is regulated only by political parties' and businesses' demand for expansion and short-term profit. We need other values and goals for activities in the Arctic. We cannot afford to play roulette with a vulnerable environment.

Poverty

Over a billion people live in extreme poverty. That means about one of every five people on earth live on less than 1.25 US dollars per day. Systems and political structures oppress the poor.

At the turn of the millennium, many world leaders stood behind eight Millennium Development Goals. This was the first time all nations agreed on common goals for global efforts to counteract poverty. The goals highlight aspects relevant for lifting people out of poverty, such as education, health, equality, and sustainable environment.

By signing the Millennium Development Goals, world leaders acknowledge that it is possible to do something about poverty. The goals breathe optimism for the future; but believing in cooperation is also pivotal for success. Only a broad, integrated approach will reduce the number who live in poverty, and improve living conditions for the world's most vulnerable people.

The objective must be to lift people out of poverty. But here is the dilemma: as millions of people gradually attain a better standard of living, greenhouse gas emissions and pollution will also increase. Thus it is crucial that privileged nations put greater effort into measures against pollution and emissions, and also relinquish some of their standard of living, to help lift people from abject poverty.

The question is whether any of the today's political leaders are willing to risk unpopularity by enacting strict environmental regulations and limiting our standard of living. The question is also whether we, as voters, will support politicians who enforce strict environmental policies.

Climate change is about rising temperatures, but also about our daily bread. Minor changes here in the north may mean devastated crops in the south. The fish in the sea are vulnerable to ecotoxins transported in currents. Minor changes in temperature may shift breeding grounds and perturb the ecological balance.

We need agreements that prevent major changes in climate and ensure global justice. We need to protect the arctic environment as a world heritage to pass along to our children and grandchildren. We need international solidarity that will lift people out of poverty and prevent climate change from making them even poorer.

Reaching binding international agreements is a slow, laborious process. But national and local agreements and actions are equally important. Unless we start small, in our local communities, there will be no change. Unless individual citizens put pressure on politicians and authorities, change will not come until it is too late. Local involvement is the key to worldwide reform.

There are countless examples of how market forces impede poor people's struggle for life and dignity. And in the struggle for a sustainable environment in sensitive areas, market forces have a major say in what direction development takes. Politicians have influence and the power of decision; thus politicians have responsibilities, but business leaders and economists also play pivotal roles. Shareholders demand economic growth. CEOs are often judged exclusively in terms of the company's profits. Those who cannot deliver satisfactory figures will be thrown out by Boards that demand a profit. Oil companies, like many other major corporations, are run with an eye to the stock market and economic prognoses, which should point upward in the eternal search for more gold.

So we must ask whether listed companies have room for ethics. Are ethical values and considerations on the agenda, or are they swept away by the focus on economic gain?

Blue light

When we hold church services at Tempelfjorden, we are in one of the world's most beautiful cathedrals. We are surrounded by Sassendalen, the icy fjord, the turquoise glacier a few kilometers to the north, and the mountain itself, the one named "The Temple". Well aware that the climate is changing, and that the Arctic is especially vulnerable, I am frequently awe-struck when celebrating mass in this snowy wilderness. There is a backdrop of solemnity behind our prayers and hymns.

Above the clouds that send a few snowflakes to earth, the stars are like peepholes in the heavens. On clear days, the stars shine down on earth and light up the dark season, our long winter night. Along with the moon and the northern lights, the starlight glistens in the white snow. The polar night may be dark, but it isn't black. There is hope. There is light in the darkness.

The sun forsakes Svalbard for a couple of winter months every year. It is a remarkable season, when the sun departs and leaves the landscape in darkness, but equally astonishing when it reappears on the horizon.

There is more light per year in the Arctic than at the equator.[45] But the light is unevenly distributed over the seasons. Our long day with midnight sun lasts four months. The polar night is equally long. In the transition period, we have twilight, the season of "blue light". On January 29, the polar night ends and we have twilight whenever the sun is less than six degrees below the horizon. Blue light season lasts until February 16 when the sun goes above the horizon for the first time. By then, we already have eight more hours of light per day. The transition from dark to light is quick, almost as if someone flipped a switch. If we include twilight, the polar night lasts from October 26 to February 16.

The snow is important. It helps cast light over the dark season. It is also important in reflecting the sun's rays back toward the atmosphere in a boomerang effect. Ice and snow. White reflects the sun, while open surfaces on sea and land absorb sunlight. When the ice melts, the open water surface grows, making the ice melt even faster.

Nature's balance is threatened. While people argue about what causes climate change, and struggle to negotiate agreements, ice is turning to water. Nature's delicate balance is approaching a point where the wounds will go too deep to heal. Areas of open sea are growing, and with them, warm areas. Nature has a limit. If we go beyond that limit, there is a risk that the negative effects will accelerate and leave us unable to turn back.

There is a lot we don't know. Even in our own time we don't have all the answers. We don't know where nature's limits are; we don't understand all the causes; we aren't aware of all the consequences. But we sense the drama taking place outside our living room windows.

As pastor of the world's northernmost church, I don't have all the answers. I know there is always more to learn. Life is more. There's something divine about it. There is more between heaven and earth than I can imagine. Scientists don't know everything. Politicians don't know everything. The church doesn't know everything. No human being knows everything. The challenge lies in how we use the knowledge we have, what we do with our lives and our community. Equally important is how we relate to what we don't know. We can choose to ignore uncertainties, or we can take them seriously.

We must seek knowledge by daring to ask questions and search for answers. We must do this for the community we live in and for our descendants. If we are to create peace between peoples, knowledge and truth are the keys. If we want a planet to live on, one that will sustain all kinds of people, we must be far-sighted and take the consequences of the questions and answers we have.

Life in the Arctic is vulnerable. The line between life and death is often thin. This is a reality for those of us who have chosen to live here in the north. It is a reality for life in nature too. Humanity and nature are parts of a single whole. We are vulnerable, but we are also strong. We can change the world. There is still time.

The ice is melting

sea ice is becoming open water
glaciers are retreating
temperatures are rising on land and sea
our course is set to the wrong harbor
the Arctic is warm
the Arctic is warm

Refrain: Give us the will to see
courage to take action
strength to play our part
Let us build a future
instead of laying to waste
we owe this to our children

ice is melting in the north
climate is changing on earth
waves crash over land, deserts and floods
our children must toil where we reaped
humanity leaves tracks
humanity leaves tracks

Earth has a fever
the Arctic is a barometer
there is hope, while the hourglass runs
we can help save lives
what we do shapes the future
what we do shapes the future

Lyrics by Leif Magne Helgesen

Gravneset (Grave Point) in Magdalenefjorden as portrayed by François-Auguste Biard, the artist on the Recherche *expedition, 1839*

The threat to the past

by Marit Anne Hauan, Director, Tromsø Museum and
Tora Hultgreen, Director, Svalbard Museum

Climate change in the Arctic in relation to graves from Svalbard's whaling era

The coast of Svalbard is isolated, cold, wild, and desolate, hemmed in by high peaks. If we go ashore for a walk, chances are we will suddenly come face to face with a whaler's grave, with skull, bones, and remnants of skin, hair, clothing, and shoes lying exposed on the ground. Archaeologist Sven B. Albrethsen calls these graves "a gloomy monument" to all the hunters who came to Svalbard to earn their fortune, but never had a chance to enjoy the results of their hard work (1987:27–28).

Many a hunter lies buried in Svalbard's barren soil, and it is not easy to tell at a glance what era the grave is from. It might be a whaler, a Russian hunter from the White Sea, or even a Norwegian hunter from more recent times. Huntsmen's graves are the most common historic relics in Svalbard. Nearly a thousand individual graves from the whaling era (1600s and 1700s) have been registered. These fascinating burial places provide unique information about European cultural history. Several of the graves have been studied archaeologically, yielding precious and unprecedented information about "peasant" clothing and how ordinary people dressed in Europe in the 17th and 18th centuries.

The whaler's profession exposed him to great risks: death was a constant companion. Whales were captured with harpoons thrown by hand from open boats; an incautious maneuver or a slap of the whale's flukes was all it took to capsize the boat or smash it to pieces. Few whalers could swim, so many drowned. And there was scurvy – "the eternal scourge of the seaman" – a disease caused by poor food and vitamin C deficiency, which threatened everyone involved in whaling voyages. Mortality was high among whalers, so planks for making coffins were an important part of the cargo on whaling vessels bound for Svalbard. Bluntly put, whalers were so certain that many of their number would die before the end of the season that they took along everything they would need for burials.

Northwestern Svalbard soon became a focal point for European whale hunting, and has the greatest density of historic remains related to this activity. It is also in this region we find the largest burial grounds from the 17th and 18th centuries, in places like Gravneset, Likneset, Ytre Norskøya and Smeerenburg. Each site holds as many as several hundred graves (Hauqebord 2010:59–68).

The whale hunter graves of Svalbard contain unique relics. This rare historic material is threatened by escalating climate change, which will have devastating effects on this important category of historic remains. Climate change poses challenges concerning how we care for these graves. We need constant awareness of the ethical dimensions.

For several years, the Directorate for Cultural Heritage has been pointing out the effects of climate change on cultural heritage sites, stating that "nature will reclaim them, aided by thawing permafrost and wilder, wetter weather".[46] The Research Council of Norway also points out that "climate change poses major threats to the cultural heritage sites in polar regions" (2013:19).

Just as climate change and melting ice will be seen first in the far north, we can already see that our cultural heritage is under growing pressure. Higher temperatures mean thawing permafrost.

The Governor of Svalbard mentions the problems posed by climate change in the new Cultural Heritage Plan for Svalbard 2013–2023. Six dimensions of climate change are highlighted as threats to historic sites: stronger storms can destroy buildings and other structures; thawing in the uppermost permafrost layer allows organic material to decompose; altered permafrost depth perturbs constructions with fundaments in the frozen ground; rot and rust spread in a moister climate; solifluction damages buildings and ruins; and increased precipitation and higher temperatures enhance erosion, which in turn threatens sites near the coastline. Higher temperatures mean less sea ice. Less ice means more waves, which also speed up erosion along beaches. The plan emphasizes that Svalbard's natural environment, and thus also the terrain surrounding cultural heritage sites, is extremely sensitive to traffic (2013:11–12).

Conditions for preservation of historic remains are changing most – and most rapidly – in this archipelago. Change affects all cultural relics. Increased precipitation damages historic sites in mainland Norway as well. For example, World Heritage treasures such as the Rock Art of Alta and Bryggen in Bergen have already been damaged because of increased water seepage and rising sea level.

The graves of the whalers are especially vulnerable. These historic relics make an incomparable contribution to European cultural history, and now this information about a bygone era is threatened. When historic sites are destroyed, we lose insight into our past: this is an ethical dilemma. Climate change hastens decomposition of historic remains, and they are being lost at an unprecedented rate.

Research ethics

Norway has had research ethics review boards for nearly thirty years. The duties of these boards are founded on the law regulating evaluation of ethics and integrity in research – the research ethics law. The committees are independent bodies within the Ministry of Education and Research. They have established guidelines intended to help researchers reflect on ethical issues. The guidelines of the National Committee for Research Ethics in the Social Sciences and the Humanities include a section entitled "Respect for human remains".[47] This section stresses the importance of researchers treating graves and human remains with respect. It is now generally accepted that graves and human remains should be treated with reverence regardless of context, and this section should therefore apply far beyond its specified context.

However, the section was not comprehensive enough to cover dilemmas encountered in research on human remains, so a special subcommittee was created: the *National Committee for Research Ethics on Human Remains* (the Skeleton Committee). The term "human remains" applies to intact skeletons, parts of skeletons, cremation ash, and other biological material.

The term "research" must be interpreted broadly, to include research-related education, outreach, and exhibits. In the Skeleton Committee's ethical guidelines, respect for the dead is the central principle.[48] Respect involves discretion, equal treatment regardless of provenance, and respect for descendants. Moreover, the guidelines emphasize the need to keep in mind the material's singularity, the feasibility of the research, the context in which it was found, and the obligation to obey laws and regulations.

The guidelines provide a particularly useful framework for ethical reflections about one group of cultural relics in Svalbard: the graves. Ethics in relation to the whaler graves is an important concern. The question is whether our research and management practices safeguard these graves in an ethically acceptable way in the face of climate change. There are important questions to ask concerning how we gather new knowledge while maintaining focus on respect for the dead.

Tragic fates

Many visitors in historic times were shaken by the condition of the graves in Svalbard, and expressed both concern and dismay. One of these was the French writer Léonie d'Aunet. At the tender age of 19 she participated in the French research cruise with the vessel *La Recherche*, which sailed to Svalbard in the summer of 1839. She is described as the first woman to set foot on Spitsbergen, and summarized her experiences and impressions from the journey in the book *Voyage d'une femme au Spitzberg*, which was published in Paris in 1854, fifteen years after the expedition. She describes her visit to Gravneset in Magdalene-fjorden thus:

> The coffins still lay on the snow, half-open and empty, probably desecrated by rapacious polar bears. It had not been possible to bury the coffins because of the thick ice; there were just a few heavy boulders on the lid to protect the dead from the wild animals. But bears are strong and had managed to move the boulders. Round about on the snow lay a few half-eaten bits of bone – a gruesome sight. With reverence, I gathered them together and returned them to the open coffins. Other coffins had been left alone and remained untouched with their ghastly content – corpse or skeleton, depending on how long they had lain there. Most were without a name, but a few had the carved inscription: Dortrecht Holland 1783 – preceded by an illegible name. A seaman from Bremen, who died in 1697. Two of the coffins lay concealed beneath a rocky outcrop and were completely intact. The dead men were exceedingly well preserved, even their clothing was intact, but nowhere did it say who they were, where they came from or when they

> died. In that desolate grave site I counted 52 coffins without headstone, without inscription, without wreath or blossom, without a mourner to shed a tear or say a prayer, without a friend to miss the departed or visit his frozen, barren resting place, where only the wild howls of storms and polar bears break the awesome silence. I was overwhelmed with panic at the thought that I too might suffer the same fate. (d'Aunet: 1968:91–92)

Her reflections and concern revolve around the buried seamen at Gravneset in beautiful Magdalenefjorden. Today the location is still dominated by the extensive burial ground situated on a small ridge above the beach; Gravneset contains between 140 and 150 graves from the whaling years. Historic sources show that the English had an important shore base for trying whale oil here, called "Trinity Harbour". In the quote, d'Aunet touches on the dilemma posed by the icy ground when the whale hunters were to be buried. Because of the permafrost, it was impossible to dig deep graves, and burials were difficult. Leonie d'Aunet also mentions that even though some of the graves were very old, the dead men were well preserved and their clothing was intact. She expresses heartfelt anguish at the condition and situation of the corpses, as well as her own dread of being left in an unmarked grave, never to be visited by friends and family.

"…fulfilling a work of love"

Others who visited Svalbard in the centuries after the whaling around Svalbard ceased also wrote of the sadness and gloom they felt when confronted with the condition of the large burial grounds along the coast of Svalbard. In 1878 a Dutch frigate anchored near the well-known Dutch whaling station "Smeerenburg" on Amsterdamøya. The crew on board were suffused with sorrow and melancholy when they gazed out over the remains of the fabled Dutch whale-oil town, and their captain said:

> One must imagine a plain, white with snow, which has melted at the water's edge, where the ground is strewn with broken red tiles and rubbish, enormous bones of whales, oars, half-rotten rope and here and there a grave… The burial-place, at the northern end of the beach had if possible a still more melancholy look, the crosses fallen, skulls and bones scattered about… (Conway 1906:180)

Out of respect for the dead, the crew erected a monument with a commemorative plaque at the grave field's highest point. In a eulogy, the dead whalers were honored for their toil "for in days gone by they did much for the honor and prosperity of our dear Country" (Conway 1906:181). Captain Beynen went on to say "it was a strange sight to behold these fourteen sturdy seamen standing at the burial-place of Dutch sailors long passed away on this distant shore. And fulfilling a work of love" (Conway 1906:181).

In unconsecrated ground?

The graves from the whaling era are between 300 and 400 years old. Some have been dug up; many have been lost to erosion. Many others have been disturbed by permafrost heave, and exposed bones have been dragged around or carried off by animals, by weather and wind, or by human visitors in search of a trophy. Nonetheless, many of the 2000 graves remain more or less intact in the arctic landscape. The people who died and were buried in Svalbard were mainly Catholics and Lutherans. They came from Europe: France, the Netherlands, Portugal, Spain and England. There were also a few Danish-Norwegian whalers. It is apparent that great care and painstaking effort were devoted to the whalers' coffins and graves – but that does not change the fact that the graves are in unconsecrated ground. The grave fields are now registered as cultural heritage sites. Some graves have been excavated and their contents – clothing, gear, and bones – acknowledged as a unique source of information about European cultural history.

In our own time, we value being able to visit a loved one's grave. A grave in consecrated ground is the last stop in our lives (Alver 1994:85). After shipwrecks when crew members are lost at sea, and accidents when bodies are not recovered, such as the capsize of the Alexander Kielland rig, the next-of-kin initiate extensive search operations to find the bodies and ensure that they can be buried in consecrated ground. Often the lack of a body prompts the bereaved to "consecrate" the ocean or the site of a catastrophe through improvised rituals of their own, such as throwing wreaths or flowers on the water or lighting candles.

The whalers' burial sites are neither consecrated nor included in rituals; all they have are commemorative plaques and cultural heritage markers. How can this be? For one thing, attitudes to death have changed considerably since the whaling era. In his book *Western*

Attitudes toward Death (1974) Philippe Ariès avers that our views about death were transformed in the 18th and 19th centuries, owing to a sharper focus on the individual. The shift meant that people's ideas about death became intertwined with fear of losing loved ones, and that death came to mean a temporary separation. Discourse about death adopted a new vocabulary, says Ariès, including expression of grief, loss, and longing. And this changed the perception of the grave. The graveyard became a place of ritual, a garden of death that gave the living a place for remembrance, a place to show others how dearly the departed was loved (Ariès 1974; Alver 1994).

This new perception of the grave as a place to remember and honor the individual does not apply to the graves along Svalbard's shores, perhaps because of the distance in time and space. The whalers' descendants may have forgotten that they have ancestors buried in Svalbard; the distance may have made the graves appear beyond reach, outside civilization. In one way, this dilemma allows us to view the graves as mere objects – cultural remains rather than shrines to the dead. Responsibility has been transferred to those in charge of preserving cultural heritage. Is that ethically defensible?

Whaling – the start of Svalbard's commercial history

The earliest commercial history of Svalbard is all about western European whaling. Willem Barentsz and his crew were in the far northwestern fringe of the archipelago when they caught sight of new land, on June 17, 1596. Their discovery can be blamed on a navigational error that proved beneficial in later years. Henry Hudson noted in 1607 that whales, walruses and seals were unusually plentiful in the coastal waters and fjords of Svalbard. This was a crucial factor in attracting thousands of European whalers to Svalbard from 1611 and into the 1700s. England, Holland, France, Spain, Germany and Denmark/Norway were particularly active at that time. Svalbard was the site of intense activity on land and at sea.

Graves from Svalbard's whaling era (1600–1800) are usually clustered in large fields, but some are in small groups. The reason we find the most extensive grave fields in northwestern Svalbard is that this was the main area for capturing the mighty bowhead whale. The first whalers came from England in 1611. Whaling operations were organized by large European companies, which established many large blubber trying stations along the shores of Svalbard.

The whale blubber was taken ashore and processed in huge trying cauldrons, where it was melted down to form whale oil. This oil was in great demand for use in lamps, for production of soap, for leather-making, and in paint. Eventually, baleen also became a prized product, used as "whalebone" in corsets, a crucial article of clothing in women's fashions of the time. The English polar historian Martin Conway emphasizes the importance of whale oil in the production of the soft, exclusive soap that was used to wash "fine clothing", writing (1906:139–140), "Perhaps it was the increased supply of good soap resulting from the discovery of Spitsbergen, that led to the great development in lace and linen which marks the costume of the wealthy at this period."

Whaling was exceedingly lucrative, and in the peak years of the 1600s, as many as 20,000 men and 300 ships would participate in the annual summer hunt along the coast and in the waters north of Svalbard. At the end of the 17th century, the Dutch whaling fleet alone comprised 150–250 ships, which captured between 750 and 1250 whales per year. This colossal slaughter meant that by the beginning of the 18th century, the bowhead whale had disappeared: along the coast of Svalbard it was essentially extinct.

Land stations for extraction of whale oil were built in several places along the west coasts of Spitsbergen and Edgeøya. In several of these, houses and workshops were erected, in addition to the tryworks. In the early whaling years, blubber was brought ashore and whale oil was extracted in vast trying cauldrons. The copper cauldrons stood in brick ovens that were used year after year. Adjacent to the land stations, burial sites were set aside. Since whaling was done from small boats, it was a risky venture. Loss of life was an everyday occurrence in the whaling industry.

That some members of the whaling crew would meet their deaths during the summer season in Svalbard was so inevitable that the expedition's supplies always included the coffin boards, sawdust and moss that would be required to give dead comrades a decent funeral. The slaughter of whales along the coasts of Svalbard meant that toward the end of the 17th century, hunters were forced to go farther out in the open sea and along the ice edge to find enough whales to make a reasonable profit. At this point, whaling activities shifted to independent whaling vessels and the shore stations fell into disuse. Whale oil was now extracted on board the vessels themselves, or the blubber was packed in barrels and tried when the ships returned home. However, the old land stations were still used, mainly to bury all the whalers who died during the hunting season.

The archeological sources

The archeologist Sven B. Albrethsen (1987:26) estimates that although only a couple of thousand graves from the whaling era remain, the total number was much higher, as we know that many graves have been "swallowed" by the ocean. Whalers' graves are generally located on low ridges not far from the old trying stations, where loose moraine made it possible to dig a grave in the summer when the sun had thawed the active layer above the permafrost. A common feature is that the grave fields lie on slopes facing the sea, unprotected from storms and waves that grind down and wash away the beach's gravel.

Archeological excavations show that the graves themselves are very simple. Uppermost was a layer of rocks. Beneath them, the dead man was laid in a custom-built coffin of planks, buried as deep in the ground as the permafrost allowed. In line with good Christian tradition, most were buried with the head to the west. Each grave was originally marked at the head end with a wooden cross inscribed with the dead man's name, year of death, and place of origin. These crosses have not survived to the present, but travelers from old times described wooden crosses on every grave (Conway 1906).

A few coffins were lined with thin linen, and some of them were decorated with lines and ribbons. The dead man was usually laid to rest fully clothed, well wrapped in woolen blankets, his head resting on a pillow of feathers or down. The remains show that the whalers were influenced by the Spanish fashions of the day, wearing wide, short breeches made of wool, short uniform-like jackets, long, thin woolen stockings with woven garters in bright colors, and low leather shoes. Under the jacket they wore shirts of linen or cotton, usually woven in striped or checked patterns. Almost without exception, the dead man was buried wearing headgear, a hat or a cap, often pulled far down over his forehead. A few had a red or yellow silk scarf around their neck.

The amount of clothing can vary, however. Not all were buried with outdoor clothing such as trousers and coats. Although most of the dead men were dressed in a shirt, breeches and long woolen stockings, a few had been laid in the coffin wearing nothing but a shirt, possibly stockings and headgear. Warm clothing and outerwear were completely absent. This might at first be interpreted to mean that such clothing did not exist – that the whalers had left home wearing their ordinary winter clothing, not in clothing especially designed for hunting whales in a far harsher climate than they were accustomed to; this view is presented by the archeologist Kristin Prestfold (2001:18). Another explanation for the lack of thick clothing, more suitable than a short wool jacket, could be that outerwear was precious, and could either be reused by other whalers, or be taken home and sold to help support the dead man's heirs. The same may be true of shoes; archeologists rarely found shoes in the graves. Conversely, headgear in the form of a knitted cap or a hat was found in the grave of every whaler. The knitted caps often had brightly colored stripes, wide or narrow. These caps and hats probably became so intimately associated with their bearer that they could not readily be adopted by other whalers. Another explanation is that the presence of fleas and lice made the idea of taking over someone else's headgear unenticing.

On closer examination, the clothing from the whaling era shows signs of long use. Detailed study of jackets, breeches and stockings reveals an incredible amount of wear; they had been darned and patched repeatedly. Many articles of clothing had clearly been sewn from older garments. This probably says something about conditions for "peasants" in Europe at the time. Cloth for clothing was expensive and must be reused to the last shred. Another possibility is that these garments were specifically intended for use on whaling expeditions, in which case repeated darning and patching signals that the garment's owner had survived several whaling seasons. In addition, the whalers in Svalbard did not come from wealthy families. Written sources show that whalers were recruited from the peasantry and the broad masses.

A consistent feature of the whalers' graves in Svalbard is that the corpse was laid to rest with great care and respect. In several graves, the dead man was wrapped in a woolen blanket with a down pillow under his head, and tucked in with a layer of moss brought all the way from home. It is touching to see how painstakingly the man's comrades had worked to ensure that he had a "comfortable berth". We believe this shows singular concern for the departed, and the bereaved comrades' anxiety that the dead man might suffer from the cold on his way to the hereafter. In some graves, the dead man is even wearing several layers of clothing: two pairs of stockings, two pairs of breeches, and so on.

Because of Svalbard's climate, the remains in the graves are often uncommonly well preserved. In the dry, cold climate of the Arctic, organic material keeps for a long time. In addition, the permafrost means that the corpse lies frozen most of the year. But warmth in the summer will melt the surface layer, allowing melt-water to seep in, and with time the material will decompose. Water that seeps in and then freezes again causes frost heaving, which with time will lift the coffin and all its contents up to the surface. This means that the boards of the coffin will eventually rot away and the corpse will be exposed to weather, wind, and predators. It is not only textiles that are well preserved, but also the dead man himself. In many cases the skeleton is intact, still bearing remnants of skin, hair, nails, beard, and tissues.

Archeological excavations of whalers' graves in Svalbard have yielded unique information about European cultural history. Everyday clothing and textiles from the 17th and 18th centuries have rarely been retrieved through excavations on the European continent, where conditions are much worse for preservation. The material from Svalbard is one of very few sources for knowledge about how ordinary people dressed at that time. This makes it unique in Europe.

What did they die of?

Written sources and archeological excavations reveal that both the crew and the leaders of whaling expeditions were poorly prepared for the harsh conditions of life in the Arctic. We have seen that the hunters' clothing was wholly inadequate to keep out the cold, even during the summer months that they spent in Svalbard. Storms and winds would have gone straight through the thin layers of wool; no "windbreaker" from this era has ever been found. We can assume that cold was a constant companion for the whalers. Frostbite and pneumonia would probably have cut short many lives in the whale-hunting grounds.

Along with frostbite and cold, overly liberal use of brandy to keep warm was a threat to survival. This is illustrated by a poem saved for posterity in the log of a Dutch ship in 1777:

> Would you like to sail to cold Greenland
> For whale, walrus, bear and traan
> You should go easy on brandy
> Put on plenty of all sorts of clothes

This was obviously good advice, and the implied plea for moderation where brandy was concerned was probably based on personal experience. During excavations at Smeerenburg, layers of broken brandy jugs were found in one of the seven buildings: it had apparently served as a tavern in the years when there was activity on shore in "blubber town". It is easy to imagine how a cup of geneva or a sip of brandy would have given welcome relief and helped keep the frost and chill at bay. Excavations also reveal that many used tobacco. Fragments from a multitude of clay pipes, along with signs of wear on the whalers' teeth where the pipe would have been, emphasize the importance of smoking. Whalers appear to have used tobacco as a stimulant and a way to survive.

Close inspection of the skeletons from whalers' graves reveals that as many as nine of twelve whalers died of scurvy. This is a relatively large proportion, and can be entirely attributed to malnourishment owing to the monotonous diet, lacking vitamin C.

Scurvy was a dreaded disease at sea and on whaling expeditions; as early as the 16th century it was known as the sailor's scourge. When the body lacks vitamin C, connective tissue does not function properly; it cannot repair itself and gradually falls apart. Collagen in the skin, gums, ligaments and blood vessels breaks down, causing bleeding. In muscle attachments, particularly on the largest bones, blood penetrates under the periosteum, and new bone is formed. This, along with cracks in the bone and signs of heavy bleeding, makes it fairly easy to detect scurvy in skeletal remains. Scurvy usually also leads to loss of teeth. We know from written sources that the whalers' diet did not include vegetables, fruit, fresh meat, or foodstuffs made with blood – all rich sources of vitamin C. The provender on a whaling expedition was essentially salted meat, salt pork, and dried peas. The whale hunters were probably already predisposed for scurvy when they arrived in Sval-

At Likneset (Corpse Point) in Smeerenburgfjord, the old whaler graves risk sliding into the water (Photo: Snorre Haukalid / Governor of Svalbard)

bard in spring. They left home before the fruit and vegetable season began. After a winter at home with little access to fresh meat, fruit and vegetables they sailed north and lived on a monotonous diet deficient in vitamin C. The fact that they did not eat fresh whale meat substantially reduced their chances of surviving the summer in the Arctic. But a few knew of a little green plant called "scurvygrass" that grew below bird cliffs. The old whale hunters called it "salad", and on old maps of Svalbard many places include this word in their names, for instance Salatberget and Salatfjellet. Written sources also show that hunters and trappers who picked and ate this plant could survive the winter.

In addition, many died of exposure, plague or drowning. This explains why Svalbard has so many graves from the whaling era (Albrethsen 1987:27).

Ethical dilemmas in cultural heritage management

The graves in Svalbard are managed as part of Norway's cultural heritage. How should people living today relate to graves from the 17th century? Are we emotionally or morally obliged to look after them? How should we balance between apparently contradictory values; can we maintain respect for the dead while simultaneously making full use of the research potential the graves offer? We know that the slaughter of whales in the 17th century nearly led to their extinction. It has taken centuries for the blue whale population to recover to a sustainable level. The same is true of the bowhead whale. This is knowledge our time can put to good use. Harvesting resources is a balancing act: one must carefully consider the situation as a whole, to ensure sustainable harvesting of nature's bounty. There is an ethical dimension in reminding ourselves of errors we have made in the past. Our ancestors left us an impoverished future, but we can create a richer future both for humankind and for the resources around us. This illustrates the complexity of the ethical issues surrounding human remains.

With this brief history of the whalers' graves of Svalbard, we want to point out the sad fact that the graves are in danger of being totally obliterated by the changes we now see in our climate. This process of obliteration has been going on ever since the graves were first created, but it has now accelerated, and the effects on what the graves contain are more severe than ever before. We implore those in charge of preserving our cultural heritage to act quickly to set up an emergency program for systematic excavation of the graves from Svalbard's whaling era, to ensure that the unique information about European cultural history that lies hidden in these graves is not lost to future generations. Systematic excavation and experienced handling of the bones will not only secure the historic information from the graves, but also guarantee that the deceased are treated with respect, and do not suffer the indignity of being washed to sea, as is otherwise inevitable when Svalbard's climate grows warmer and wetter.

Printed sources:

Albrethsen, Svend B. 1987: Ikke bare en historie om tran. In: *Ottar* 5/87, 26–35

Alver, Bente Gullveig 1994: Det genvundne paradis, s. 66–90. In: *I dødens skygge*, Vett & Viten AS

Ariès, Philippe 1974: *Western Attitudes toward Death*. Johns Hopkins University Press, Baltimore, Maryland

Conway, Martin 1906: *No Man's Land*. University of Cambridge

d'Aunet, Léonie 1968: *En pariserinnes reise gjennom Norge til Spitsbergen anno 1838*. Aschehoug: Oslo (in Norwegian). Originally published in 1854 under the title *Voyage d'une femme au Spitzberg*

Hacquebord, Louwrens 2001: English and Dutch Whaling Stations in Spitsbergen in the 17th Century. In: *Whaling and History 111*, ed. Ringstad, J.E., 59–69

Prestvold, Kristin: *Smeerenburg, Gravneset. Europe's First Oil Adventure.* Booklet published by the Governor of Svalbard (in Norwegian)

Research Council of Norway 2013: Norwegian Polar Research. The Research Council's policy for 2014–2023 (in Norwegian)

Online sources:

http://www.miljostatus.no/Tema/Kulturminner/Klimaendringer-og-kulturminner/

http://www.etikkom.no/Aktuelt/Aktuelt/Fagbladet-Forskningsetikk/Arkiv/2011/2011-4/Nar-liket-ikke-er-blitt-skjelett/

http://www.regjeringen.no/pages/38311590/PDFS/STM201220130035000DDDPDFS.pdf

Ten countries have permanent research stations in Ny-Ålesund, and scientists from an additional ten countries also do research from the world's northernmost community, at 79°N (Photo: Eva Therese Jenssen)

Climate challenges and cooperation between nations

by Ole Arve Misund, Managing Director, University Centre in Svalbard / Chief Scientist II, Institute of Marine Research

Humanity has divided the world into nations with their own languages, cultures, religions, forms of government and ways of life. These nations are divided into states, regions, counties, cities and villages. Citizens in many countries – at least in the western world – have private residences that they maintain and improve to the best of their ability.

But the sun shines on all of us. It provides warmth, lets plants grow on land and in seas, and drives wind and weather. Sunlight is not in any way regulated by humans or nations. Solar radiation depends on the time of day, the geographic location, the longitude at which we live, and the latitude North–South.

The same is true of climate, which means *weather over time*. It varies considerably with geography and season. It is completely different in winter, spring, summer, and fall; it changes with latitude and longitude, near oceans, seas and lakes, from sea level to mountaintops. It is influenced by ocean currents and topography. It can be hot and arid, as in the deserts, but also hot and humid, as in the lush tropical rain forests.

Since finishing my graduate work in fisheries biology at the University of Bergen in the mid-1980s, I have followed both the scientific marine research literature and the scientific debate in general. In the marine research field, climate has long been a topic of interest. For graduate students at the Institute of Fisheries Biology, the draft of an article by Sætersdal and Loeng[49] was required reading. It described how climate affects recruitment to the northeast–

arctic cod population in the Barents Sea. Warm years were favorable for strong year classes, whereas cold years usually gave poor recruitment.

The first claims that the climate might be changing came in the mid-1990s. Research environments are usually critical to new hypotheses and theories, and many were initially skeptical toward observations and conclusions pointing at climate change. It took a long time before marine scientists began discussing climate *change*, but climate *variation* was well known, and the odd things happening with the climate were initially assumed to be part of the same well-known phenomenon. When we gathered in Copenhagen in 1999 to plan a strategy for the International Council for the Exploration of the Seas, the notion of including research on climate change met strong resistance, but when the plan was finalized in 2002, the topic was included. The Marine Research Institute in Bergen was also slow to focus on climate change, but in 2003 the Institute joined with others to establish the Bjerknes Centre for Climate Research, which the Research Council of Norway later raised to the status of Center of Excellence.

As a member of the Research Council's Global Change Committee, I gained insight into the IGBP program[50], which created the Global Change Index and monitored several other parameters that clearly indicated the reality of climate change. The trends are sobering, and to put ourselves in a better position to detect changes and give advice about necessary countermeasures, the scientific community has united behind a new initiative called *Future Earth*.

Climate change

Earlier changes in climate have been caused by natural conditions, but it is claimed that the changes we now see are caused by humans. Since the Industrial Revolution, the main sources of energy used for heating, transportation, and industrial processes have been fossil fuels. First coal – still the most important energy-bearer – and then a steadily increasing amount of oil and natural gas. Combustion of such fuels gives CO_2, which is essentially always released into the atmosphere. As concentrations of CO_2 gas increase, the atmosphere retains more of the solar heat reflected from the earth's surface. As a result, global temperature will increase in parallel with increasing atmospheric CO_2 concentration.

The theory is soundly based on the principles of physics and chemistry. But the consequences for humanity are overwhelming and difficult to take in. The modern lifestyle – our cars and planes always in motion, diesel-fueled vessels plying the seas, and in many countries electricity from power plants that run on coal or gas – all this will come under threat if we are no longer able to use fossil fuels. So we procrastinate, ignore the issues, and allow ourselves, deep in our hearts, to doubt the entire CO_2 theory.

Can we be sure that future developments will match what the climate models predict? Maybe we don't yet understand the interactions that shape climate well enough to create models that can predict developments reliably? And who can say anything at all about what the world will look like a century from now? Meteorologists, who work with day-to-day variations of wind and weather, are hard put to make accurate weather forecasts for the coming week. They get it wrong now and then, as we all know. Weather is important; all Norwegians are interested in the forecast because weather strongly affects which outdoor activities are possible. So, despite inaccuracies, we rely on the meteorologists nonetheless. In fact, according to Synovates opinion polls, the Norwegian Meteorological Institute is the Norwegian institution most people trust – and has been so for the past six years (2008–2013)! But it is more difficult relating to climate models and the projections they provide – at least at a personal level. We drive our cars more than we used to, and airfare has become so cheap that air traffic is burgeoning worldwide.

All over the world, people expect continued economic growth, to the benefit of themselves, their families, and society as a whole. But most of the energy used for industry, transportation and generation of electricity still comes from fossil fuels. So CO_2 emissions to the atmosphere continue to rise. A series of measurements taken at the NOAA[51] station on Mauna Loa in Hawaii, starting in the International Geophysical Year of 1957, show a steady increase from about 320 ppm[52] CO_2 to nearly 400 ppm CO_2 in 2013. In the winter of 2013, CO_2 concentrations above 400 ppm were measured in Ny-Ålesund at the Zeppelin Station, run by the Norwegian Polar Institute, Stockholm University and NILU – The Norwegian Institute for Air Research.

The United Nations' Intergovernmental Panel on Climate Change (IPCC) presented its Fifth Assessment Report in September 2013. The main conclusion was that scientists are now much more certain that the climate is actually changing, and that this is due to humanity's

Zeppelin Station in Ny-Ålesund was established in 1988–1989. The current building was completed in 2000 (Photo: Eva Therese Jenssen)

use of fossil fuels, which causes release of CO_2 to the atmosphere. The IPCC does not do research itself, but compiles and evaluates all research published on the topic, in this case up until 2013.

The UN Framework Convention on Climate Change

The United Nations has established a Framework Convention on Climate Change (UNFCCC). At present the convention has 195 parties (nations and groups of nations), giving it essentially worldwide coverage. The framework convention is bound to the Kyoto Protocol and the Doha Amendment, which will be discussed below. The ultimate goal of the convention, and the treaties that have been negotiated, is to stabilize the atmospheric content of greenhouse gases at a concentration that will prevent dangerous human-induced effects on the climate system.

The United Nations Climate Change Conferences

Once a year, political representatives from around the world meet to discuss climate issues. The conference is organized by UNFCCC and is arranged by member states in the Framework Convention. The objective of this Conference of the Parties (COP) is to review the practical implementation of the convention and other legal instruments that COP adopts. The delegates also make decisions required to facilitate or promote implementation of the framework convention, through institutional and administrative arrangements,[53] and put in place international regulations to limit harmful emissions to the atmosphere.

The first COP meeting was held in Berlin in March 1995. The COP secretariat is in Bonn, and the meetings are also held there, unless one of the parties offers to host the conference. The post of president rotates between the United Nations' five world regions: Africa, Asia, Latin America and the Caribbean, central and eastern Europe, and western Europe and others. The meeting site tends to rotate between these regions along with the presidency.

The Kyoto Protocol

Much progress has resulted from these conferences. The most well-known outcome is the protocol approved at COP3, which was held in Kyoto in December 1997.[54] Nations could sign the protocol at UN Headquarters in New York from March 1998 to March 1999, and over that time, 84 nations signed. But nations that participate in UNFCCC can ratify, sign and enter into the protocol at any time. The Kyoto Protocol came into force on February 16, 2005, i.e. 90 days after no fewer than 55 nations that were responsible for not less than 55% of the carbon emitted in 1990 from the nations belonging to the UNFCCC, had ratified, signed and entered into the protocol. At present, there are 192 signatures under the Kyoto Protocol (191 nations and one regional economic integration organization). Norway is one of the nations that have ratified the Kyoto Protocol. But many nations – including the United States – have yet to sign.

The Kyoto Protocol is an international agreement founded on the UNFCCC; it obliges the parties to set binding international goals to reduce greenhouse gas emissions. Clearly, after 150 years of industrialization, the developed countries account for the largest emission of greenhouse gases to the atmosphere, and the protocol places a greater share of the burden on these countries according to the principle "common but differentiated responsibilities". Detailed rules for implementation of the Kyoto Protocol were adopted at COP7 in Marrakesh, Morocco in 2001, with the first commitment period running from 2008 through 2012. In Doha, Qatar, in December of 2012, another amendment to the Kyoto Protocol was adopted, with new commitments and a new commitment period, additions to the list of greenhouse gases to be monitored, and updates of various sections of the protocol.

In the first commitment period, 37 industrialized nations and the EU committed to reducing greenhouse gas emissions by 5% compared to those in 1990. For the second commitment period, the partners vowed to reduce emissions to at least 18% *under* the 1990 emissions during the eight-year period from 2013 to 2020. However, the composition of the partnership differs in the two commitment periods.

Before the COP15 meeting in Copenhagen in December 2009, expectations were sky-high: an agreement would be reached that would "save the climate". Participants included some of the world's top political leaders, such as presidents Barack Obama, from the United States, and Luis Silva, from Brazil. But the results of the conference were disappointing.

The 19th Climate Conference, 2013

The 19th UN Climate Change Conference (COP19) was held in Warsaw, Poland, in November 2013. According to Norwegian media, the results were meagre, and we got the impression that no binding agreements had been reached. Environmental advocacy groups protested against what they saw as the delegates' lack of initiative by leaving the conference en masse. But they promised they would come back even stronger.

A closer look at the outcome of the Warsaw conference reveals a more nuanced picture. The nations agreed on a pathway for working toward a new universal climate agreement in 2015, and several decisions were made that will contribute to reducing greenhouse gas emissions. COP21 will be held in Paris in December 2015, and the fact that nations can submit their suggestions for emission cuts in advance has raised expectations. The potential new agreement in 2015 will prepare the way for a new commitment period starting in 2020. At the Warsaw conference, there was consensus for establishing an international mechanism to assist populations threatened by loss and damage owing to extreme weather events or to gradual changes, such as sea level rise. Governments sent clear signals of willingness to provide financial support to developing countries in efforts to cut emissions and adapt to climate change. Several countries, including Norway, Great Britain, USA, Korea, Japan, Sweden, Germany, and Finland, as well as the EU, promised financial support to developing countries for specific measures to counteract climate change. The agreements also included measures against deforestation. Forests absorb about one fifth of the greenhouse gas emitted by humans. The Warsaw framework for the REDD++[55] program led to promises of financial support totaling over one billion NOK from nations including the United States, Norway and the United Kingdom. An important milestone was that 48 of the world's poorest countries set up an extensive plan for how to handle the effects of climate change. With these plans, the nations will be better placed to assess the consequences of climate change and determine how they can improve their ability to adapt. Many countries pledged financial contributions to an adaptation fund. Groundwork was also completed for a climate technology network and center that developing countries will be able to approach for advice and assistance on using relevant technology.

Hopes for reduced emissions?

Over time, the UNFCCC, the COP meetings, and the Kyoto Protocol with the Doha Amendment have established a framework and a forum for international negotiations that raise hope for the future. During the first and second commitment periods, many nations and groups of nations signed binding agreements to reduce greenhouse gas emissions to a specific level. The conference in Warsaw in November 2013 staked out a pathway toward new binding agreements in 2015, which means that from 2020, the focus will be on real emission cuts in as many as possible of the world's nations. It is likely that the UNFCCC and the COP meetings are the only way to achieve real reductions in greenhouse gas emissions through binding international agreements.

Development of sustainable "green" technologies to replace fossil fuels will be needed if the world's nations are to reach their emission reduction goals. In Norway and other countries with many waterfalls, hydroelectric power is a safe, reliable, renewable resource. Other countries with more sun and warmth will benefit from solar energy as solar panels become cheaper, better and more reliable. Wind turbines are increasingly common in many places, on land and at sea. But they are not as efficient as had been hoped and it will take many turbines to make a dent in our energy needs. Nuclear power is another alternative, but the risk of serious accidents and the dire consequences of radioactive leakage, as from Chernobyl in 1986 and the Fukushima incident after the tsunami in Japan March 11, 2011, make many skeptical to continued investment in this energy source. For example, Germany has resolved to shut down its nuclear power plants.

"In the future, the environmental costs must be reflected in the price we pay for goods and services, regardless of sector." These are the words of Irene Waage Basili, Managing Director of CG Rieber Shipping, quoted in the daily newspaper *Bergens Tidende* December 21, 2013. She argued that regulation is required to control people's behavior. She also declared that it was unrealistic to expect a "green" economy to come about solely through individual citizens making moral choices. A shift is unlikely to happen until it is cost-effective to think green. The challenges posed by climate change cannot be resolved by market forces alone. As of January 1, 2014, the Norwegian Environment Agency stopped selling emission rights. Researchers Hans Jakob Walnung and Carlo Aall[56] even claim that the climate solutions we have at present are part of the problem, because the energy saved through efficiency measures and use of new technology is squandered in increased energy consumption.

Norway's former Prime Minister Jens Stoltenberg was appointed UN Special Envoy on Climate Change on December 23, 2013. The intention was that he should work toward achieving a binding climate agreement in Paris in December 2015. Almost immediately, questions were raised as to whether he was the right person for this prominent position. It was not the first time Jens Stoltenberg had straddled two horses simultaneously. Writing in the daily newspaper *Aftenposten*, the same day as the appointment was announced, Ole Mathismoen claimed that Stoltenberg had arranged for the quickest possible extraction of oil from Norwegian fields, while at the same time acting as a spokesperson for climate, climate science and trade in CO_2 emission quotas. On Friday, March 28, 2014, Jens Stoltenberg was appointed Secretary General of NATO. Until he took on his new role October 1, 2014, he continued as UN Special Envoy on Climate Change. On May 2, 2014, he visited the University Centre in Svalbard and gave a lecture to students, employees and guests concerning future climate challenges and how the climate negotiations have been set up to meet those challenges. He made it very clear that even though emissions all around the world have continued to increase, there is hope that the international negotiations to limit emission of greenhouse gases will be successful. New technology that makes it possible to extract more energy from sun, wind and biological sources gives hope for the future. Economic incentives to stimulate increased use of environmentally friendly technology, and higher taxes on use of fossil fuel will be required, said Jens Stoltenberg. He concluded by saying that measures to limit climate change must be implemented at the same time as measures to facilitate economic development that will lift more people out of poverty and destitution.

Like Jens Stoltenberg, Michael Bloomberg, the former mayor of New York, has been appointed Climate Envoy by UN Secretary General Ban Ki-moon. Bloomberg is to serve as Special Envoy for Cities and Climate Change, with the task of keeping in touch with mayors and other key players to mobilize political support for climate-related measures in cities. His work is part of the Secretary General's strategy to strengthen both climate actions and the UN's profile in this context.

On Palm Sunday, April 13, 2014, in Berlin, the IPCC Working Group III presented its contribution to the Fifth Assessment Report, concerning mitigation of climate change. The report points out that:[57]

- emission of greenhouse gases has not yet started to decrease
- it is not yet too late for measures that will limit global warming to 2°C
- emissions must be 40–70% lower in 2050 than in 2010
- China was responsible for most of the emissions increase from 2000 to 2010
- 80% of the electricity generated in 2050 must come from renewable energy sources
- combustion of fossil fuel without carbon capture must be phased out in this century
- energy conservation measures are vital to reduce carbon emissions before 2050
- taking immediate action will cost less than postponing climate measures until later
- international cooperation is necessary to attain the emission goals

The measures presented by the IPCC to limit climate change in the future are drastic, and require significant restructuring of the energy and transport sectors. Major technological advances are needed, and energy production must be transformed. International collaboration will be pivotal in attaining the emission reduction goals through binding agreements.

The 21st Climate Conference in Paris

The groundwork being done by high-profile politicians such as Stoltenberg and Bloomberg raises expectations for important results at COP21, which will be held in Paris in December 2015. The French government is also eager for this meeting to be a success. A central figure is Jacques Lapouge, who has been appointed Ambassador responsible for Climate Negotiations by the French Ministry of Foreign Affairs. He has extensive experience of working abroad and has been active in many places, including countries in Africa. In the summer of 2013 he participated in the High North Study Tour around Svalbard, organized by the Norwegian Ministry of Foreign Affairs, the Norwegian Polar Institute, UNIS, and SINTEF. There he heard about the effects of climate change that we see in the environment in Svalbard, and what changes scientists project, and he participated in discussions about what is required to come to grips with the changing climate. It remains to be seen whether seven hectic, informative days of camaraderie on board the research vessel *Lance* will have any effect, but at least Lapouge has a good starting point for a forward-looking conference. Time will tell what the outcome will be.

On July 3, 2014, Laurent Fabius, France's Minister of Foreign Affairs, who will host the conference in Paris in December 2015, visited Ny-Ålesund together with his Norwegian counterpart Børge Brende, to see and learn about the effects of climate change on arctic ecosystems. The following day, the Foreign Ministers visited UNIS, where our French PhD student Heidi Sevestre gave a lecture on how the glaciers in Svalbard are changing as a result of climate change, and about the research being done to understand and quantify how melting glaciers will affect sea level in coming years. The ministers Fabius and Brende went home supplied with many pertinent facts to influence negotiations in Paris in December 2015.

"Breaking the Climate Stalemate" – the Ny-Ålesund Symposium 2014

Ny-Ålesund, Svalbard's northernmost community with inhabitants year-round, is a village dedicated entirely to research. In May 2014, a symposium was held there on the topic "Breaking the Climate Stalemate".[58] Climate-related symposia have been organized in Ny-Ålesund annually since 2006; the participants are few – just 45 – but many are highly influential. I had the good fortune to be invited to participate in 2014, and heard presentations by knowledgeable speakers, some of whom are closely involved in international climate negotiations. The host of this year's symposium was Tine Sundtoft, Norwegian Minister of Climate and Environment. In her presentation, she emphasized that neglecting to prepare ourselves for climate change is not an option, and that we must start preparing immediately. Several of the presentations are available at www.ny-aalesundsymposium.no, but I would like to highlight key points in some of the most important ones.

IPCC Chairperson R.K. Pachauri went through the main results in the part of the Fifth Assessment Report prepared by Working Group III, concerning how to boost measures aimed to mitigate climate change. To keep the increase in world average air temperature under 2°C, atmospheric CO_2 concentrations must be kept under 450 ppm. This means that before 2050, emission of greenhouse gases must be cut by 40–70% compared to 2010 emissions. Energy production from low-carbon, renewable energy sources much be trebled or quadrupled; use of nuclear power and biofuels must increase, and fossil fuel combustion must be subject to carbon capture and storage by 2050. Pachauri showed that even if the most ambitious countermeasures are implemented, global consumption in 2030 would only be 1.7% under the current total world GDP; in other words, the overall cost of the climate measures needed to keep the world inhabitable is relatively small.

The former president of Mexico, Felipe Calderon, presented possible ways to achieve economic growth without detrimental effects on climate. He is leader of a UN project called "The New Climate Economy", in which Norway also participates. The objective is to ensure economic growth and a livable climate simultaneously. A central theme is more urbanization and improved infrastructure in cities, to keep carbon emissions per inhabitant as low as possible. More renewable energy, a shift from deforestation to reforestation, and technological innovation are also central methods for achieving economic growth despite drastic cuts in greenhouse gas emissions.

Climate and energy adviser Kjetil Lund stressed that climate policy entails changing economic behavior. He explained carbon pricing and cuts in fossil fuel subsidies. Carbon pricing systems are cropping up all over the world and are leading to more renewable energy, better energy efficiency, and technological innovation. Without a CO_2 surtax, emissions from the Norwegian oil industry would be considerably larger. About a third of all new cars in Norway are now either electric or hybrid. Carbon pricing is needed and has now been introduced in thirty countries with over three billion inhabitants in all. The price of carbon is still too low, but there is no doubt that pricing can be an extremely effective measure.

The Minister of Climate and Environment summed up the symposium thus:

- A new climate agreement must be reached in Paris in December 2015. It must be ambitious and give concrete numbers.
- There is broad agreement on the need for carbon pricing and reduced use of fossil energy.
- We need a new partnership with market and economic forces.

Will the major nations lead the way?

On July 2, 2014, we heard that the United States proposed to reduce CO_2 emissions from existing power plants by 30% before 2030. Each individual state would be allowed to determine how to achieve that 30% reduction. But emissions from power plants only comprise about 40% of the country's CO_2 emissions, and since the emission cut is to be in relation to emission levels in 2005, and emissions have already fallen by 10% because of transition from coal to shale gas, the US is already well on the way to achieving this goal. The proposal came from Gina McCarthy, Administrator of the Environmental Protection Agency, who emphasized that by using cleaner energy sources and cutting energy waste, the plan would ensure cleaner air and help prevent climate change. She described it as "an investment in better health and a better future for our kids." These are new signals from the US government and may send an important message to other major carbon-emitting nations that something must be done. Nevertheless, the proposal is controversial in the US, particularly among Republicans, many of whom still deny the existence of human-induced climate change. Still, the proposal shows that the US government recognizes the seriousness of the most recent IPCC reports stating that the climate changes we now observe are likely to have been caused by humans – something no US government has previously done. Colleagues in positions of leadership at NOAA described how they, at the annual meeting of ICES (International Council for Exploration of the Seas) in Halifax in 2008 during the Bush administration, were not allowed to write about climate change in their reports and publications, but must limit themselves to the term climate *variations* (and this in the country that prides itself on the constitutional right to free speech!). But now it looks as though the United States is moving in the right direction on climate issues. That will probably have a decisive effect on international negotiations for binding international climate agreements, and not least on technological developments aimed to find new energy sources with lower emissions. In a speech in May 2014, President Barack Obama explained that American businesses, and state and federal powers should invest in solar energy technology to replace polluting energy sources.

Results of international cooperation in other environmental contexts

There are countless examples of how international cooperation to solve environmental issues can lead to success. But sometimes results take time. The international cooperation I know best involves efforts to ensure sustainable fisheries. Fish populations often cross national borders in the sea, and international cooperation is needed in both research and management, to establish regulations that allow sustainable harvest of these populations. After more than 40 years of giving advice concerning fisheries, ICES announced[59] in Copenhagen recently that exploitation of 85 of the most important fish populations in the northeast Atlantic had declined substantially in the past decade; this was seen as an important step toward sustainable fisheries in the future. The cod populations around Iceland, in the Barents Sea, and in the Baltic; the herring populations in the North and Norwegian Seas; and the flounder population in the North Sea were being harvested less, and the fisheries on these populations were now sustainable. This was achieved through quality-assured scientific advice and cooperation between the nations in the Northeast Atlantic Fisheries Commission (NEAFC), along with multi- and bilateral negotiations between coastal states such as Norway and Russia. These two countries manage the fish populations of the Barents Sea under the auspices of the Joint Norwegian–Russian Fisheries Commission, established in 1976.

But cooperation in fisheries research between Norway and Russia/USSR goes back even farther. In 1957 scientists from Russia's Polar Research Institute of Marine Fisheries and Oceanography (PINRO) in Murmansk visited the Institute of Marine Research in Bergen; they were concerned that the large-scale industrial trawling that had developed in the Barents Sea after the Second World War would harm the northeast–arctic population of cod. Norwegian marine scientists visited Murmansk the following year, and this was the beginning of an extensive fisheries research collaboration between Norway and the Soviet Union, which now continues between Norway and Russia.[60] Starting in the early 1960s, coordinated research cruises were undertaken to assess the number of young fish (fry) in new year-classes in the Barents Sea. Starting in 1974, there were coordinated censuses of capelin, and from the winter of 1980, coordinated winter cruises to census cod, haddock and other deep-water fish in the Barents Sea. These studies formed the basis for scientifically based advice about quotas, and the autumn meeting of the Joint Norwegian–Russian Fisheries Commission set the quotas for the following year.

Despite the availability of research-based data on population size and estimates of sustainable harvest, and despite the existence of a framework for negotiating national fishing quotas, overfishing continued in the Barents Sea through the 1980s and 1990s.[61] Around the turn of the millennium, authorities still exposed considerable illegal fishing, in excess of the national quotas set by the Joint Norwegian–Russian Fisheries Commission. The Norwegian

UN Secretary-General Ban Ki-moon visited Ny-Ålesund in 2009 to learn about climate change in the Arctic (Photo: Max König / Norwegian Polar Institute)

Coast Guard and the Directorate of Fisheries put much effort into tracking down and controlling this illegal activity. It was not until 2005 that the harvest of cod in the Barents Sea fell to levels considered sustainable over time, but in combination with favorable natural conditions, this laid the foundation for strong yearly recruitment and significant growth of fish populations. In the past two years (2013 and 2014) the quotas for northeast–arctic cod have been record high: one million tons per year.

The future of the Arctic

In the same way as the challenges of climate change emerged gradually over time, and many climate negotiations have been held in attempts to face them, it has taken time to deal with other, simpler environmental challenges, such as overfishing. But scientifically based knowledge, expert input, and formal, painstaking international deliberations have enabled us to take control of environmental problems and lead developments in a positive direction, as shown by the case of overfishing.

Given our current knowledge of climate and projected climate change, and the existence of an international framework for climate negotiations, there is reason to believe in solutions and measures that will enable us to limit greenhouse gas emissions enough, so that we can prevent drastic global warming before the end of this century. We may even manage to limit the warming to about 2°C. Our modern, energy-intense lifestyle, with its massive consumption of fossil fuels, and the greenhouse gas emissions it entails, mean that humans now influence what happens on the earth's surface and in the atmosphere. So strong is this effect that scientists are calling our time *the Anthropocene Era*.

In many regions, a 2°C rise in average temperature will have devastating effects: reduced food production and deteriorating living conditions. The Arctic will see major changes: less ice and worse living conditions for animals adapted to this environment. Around Svalbard, we expect more open water and more fish; the terrestrial fauna will undoubtedly also change. But conditions for those who live here may actually improve, even as they deteriorate elsewhere on the planet.

Many cherish a hope that the expansion of ice-free areas and a milder climate will allow extraction of more energy resources from the Arctic. Plans are being made to secure knowledge and create new technologies that will make it possible to extract oil and gas from the Arctic without harming the environment. But the cost of such innovation may be very high.

In the perceptive book *After the Ice*,[62] author Alun Anderson writes:

> The Arctic's message is finally getting through to the wider world. The ice is melting away and it is shouting, "Your planet is in danger." A price is going to be put on carbon emissions and that will drive the search for alternatives to oil. My bet is that changing energy use will make all but the easiest to find arctic oil too expensive in the next few decades.

Nonetheless, there is reason to believe that shrinking ice and milder climate in the Arctic will open up larger areas of the ocean for fishing, shipping, and petroleum operations. In our part of the Arctic, we have an obligation to manage these challenges responsibly.

In Svalbard, management plans are being developed for nature preserves in the eastern[63] and western[64] parts of the archipelago. All told, these two nature preserves cover about 60% of Svalbard's land area, and certain types of human activity will be strictly regulated. It is unlikely that there will be any major industrial operations apart from the mines already run by Store Norske Spitsbergen Kulkompani A/S in Longyearbyen–Svea and Trust Arktikugol in Barentsburg. But research, education, travel, and tourism can probably continue to expand in Svalbard.

The ocean management plan developed over the past decade, covering waters from parts of the Barents Sea in the north to the North Sea in the south, provides a formal framework and a system that can be used to weigh nature conservation and environmental considerations against accessibility of areas for fishing, shipping, and oil and gas operations. With this plan as a foundation, Norwegian authorities have opened up for petroleum operations in selected areas as far north as 74°30'N, at the latitude of Bjørnøya. Important breeding grounds for the major cod populations of Lofoten–Vesterålen–Troms II have not yet been opened, nor have the Møre sectors off Ålesund–Molde, which are critical breeding grounds for Norwegian spring-spawning herring. So far, the management plan for the Barents Sea and Lofoten (first laid out in 2006 and revised in 2011) only covers areas south of 74°30'N. But if an increasing amount of ocean becomes ice-free more regularly, and the climate becomes milder in the north, the plan may well be extended to waters even farther north.

The management plan system, developed for both land and sea, allows government agencies to make decisions concerning various types of exploitation of new areas in the north in a sound, science-based, responsible way.

Adventfjorden and Hiorthfjellet on a bright spring day (Photo: Eva Therese Jenssen)

Responsibility, options and choices

by Kim Holmén, International Director, Norwegian Polar Institute

Climate is changing, has always been changing and will always change. Humans have survived through all past changes for as long as we have existed. We know little about the ordeals or opportunities that these changes (e.g. the ice age cycles) bestowed on our forefathers, but we know that the changes were caused by influences other than those of the human beings themselves. We also know that there have been enormous changes in climate and atmospheric conditions with much higher carbon dioxide concentrations than those of today – but those changes were before humans existed. What changes life as such has survived on earth, need not be relevant for what changes humans and society can endure.

Humans have during the past few centuries become a powerful force on earth. We impose changes virtually everywhere, through hunting, farming, forestry, fertilization, harvesting, mining, and war. We make buildings, roads, noise, artificial light, magnetic fields, and radioactive wastes. We divert watercourses, emit chemicals, dump wastes, alter erosion, move material, relocate species, manipulate genetics, alter the composition of the atmosphere and oceans, influence climate and more. Most of these changes are performed with good intent of providing humans with a better life. Many of the changes, however, also create negative effects. Frequently the negative effects appear slowly or are difficult to pinpoint in a one-to-one cause and effect relationship. There is often a disconnection in time and/or space between the desired positive outcomes and whatever negative consequences there might be.

Which of the human-induced changes are most serious? During the past decades the environmental debate has focused on climate change as the most serious environmental con-

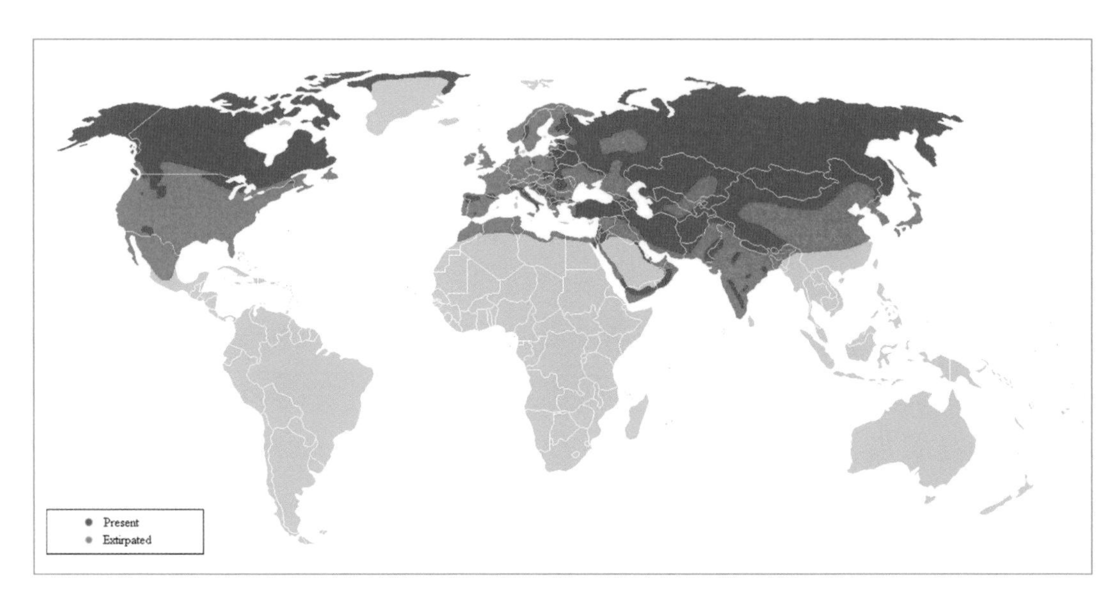

The past and present range of the gray wolf (Illustration: Wikipedia Commons)

cern of our times. In the Arctic, climate change is clearly visible today and this is the region where the largest and most rapid changes are expected.

The map above shows the present range of gray wolves and the areas where wolves roamed before human intervention. These changes are indeed dramatic, imposed by man, but not through climate change. The wolf's original range is evidence of a species with extraordinary adaptability to climate, with presence in places as disparate as northeastern Greenland, the deserts of Mexico, the Arabian Peninsula, and the tropical forests of South India. It is likely that the wolf would be robust in most of its range also under the most extreme scenarios of human-induced climate change. Yet the wolf has been extirpated from large parts of its original range due to other conflicts with humans, mainly for protection of livestock and safety of people. Even in the affluence of today's Scandinavia, where the loss of some livestock can readily be compensated and everyone has safe housing, strongly voiced opinions call for eliminating the miniscule number of wolves that remain.

The changes in the range of the wolf are large also compared to predicted climate change, for example model calculations of permafrost thawing (illustrated next page) towards the year 2100. The changes imposed on the wolf are generally accepted and seldom discussed as "drastic", whilst a movement of the thawing line a few hundred kilometers northward is frequently painted as a "catastrophic" consequence of human-induced climate change. Both views need to be nuanced, but even then they both remain serious perturbations created by humankind.

The Arctic has special dimensions that make climate change particularly significant in this region. Many species are highly specialized and live at a terminus; they have nowhere to escape if the habitat changes and no frigid areas remain. The map on page 234 shows the present distribution of the narwhal. In the context of human-induced climate change and other human perturbations, the narwhal is in many ways the opposite to the wolf in having an extremely narrow climate band that provides for it. The narwhal is dependent on sea ice

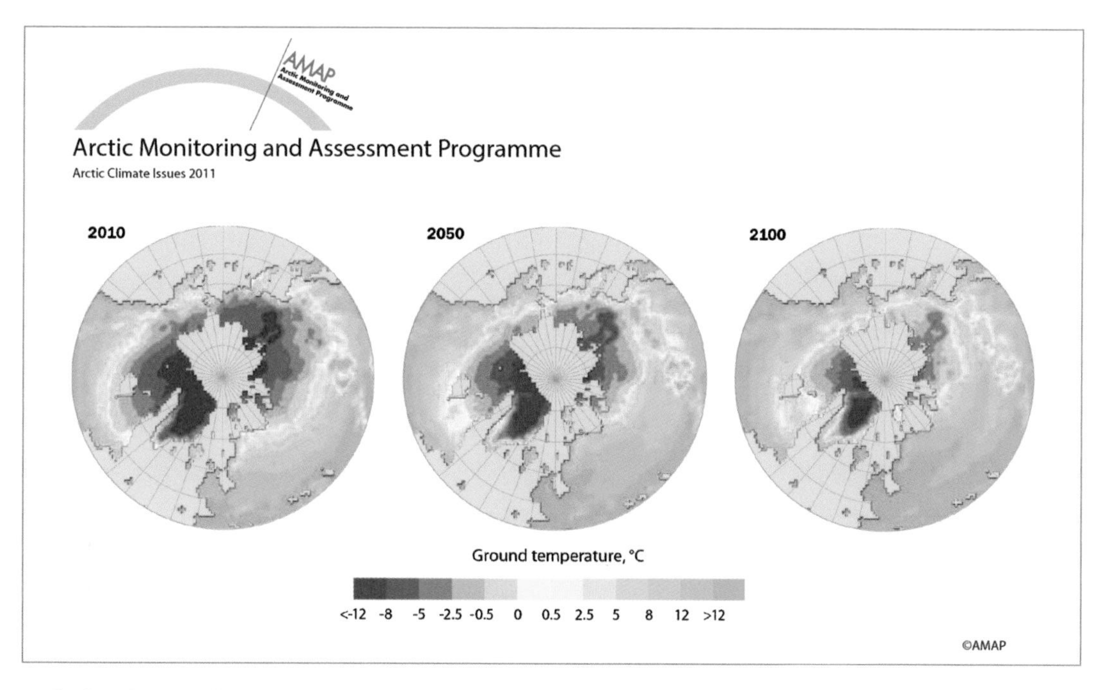

Calculated permafrost thawing up to year 2100 (Illustration: Arctic Monitoring and Assessment Programme)

and the ice edge for protection and food. Contrary to the wolf, it has few direct conflicts with humans even though there is subsistence hunting of narwhal in some of its range. The narwhal is extremely sensitive to underwater noise (e.g. the churning of propellers), but because it lives in ice-covered areas, that conflict has had limited relevance in most of its range, although Svalbard may be an exception. The largest human threats for narwhals arise from climate change and pollutants.

The narwhal is particularly vulnerable to climate change because of its restricted geographic range and its highly specialized diet. Narwhals are a clear example of why the effects of human-induced climate change deserve special attention in the Arctic. Climate change is not caused by human activities in the Arctic, yet this is a region that already is experiencing large changes. Narwhals, polar bears, the ivory gull, and most other specialized high arctic species are rapidly losing habitat.

Most people in the world will never experience a narwhal or polar bear in the wild, but many would, nevertheless, consider the world poorer without them. It is an ethical issue whether our actions should allow the pressures on the Arctic to increase. Humans have

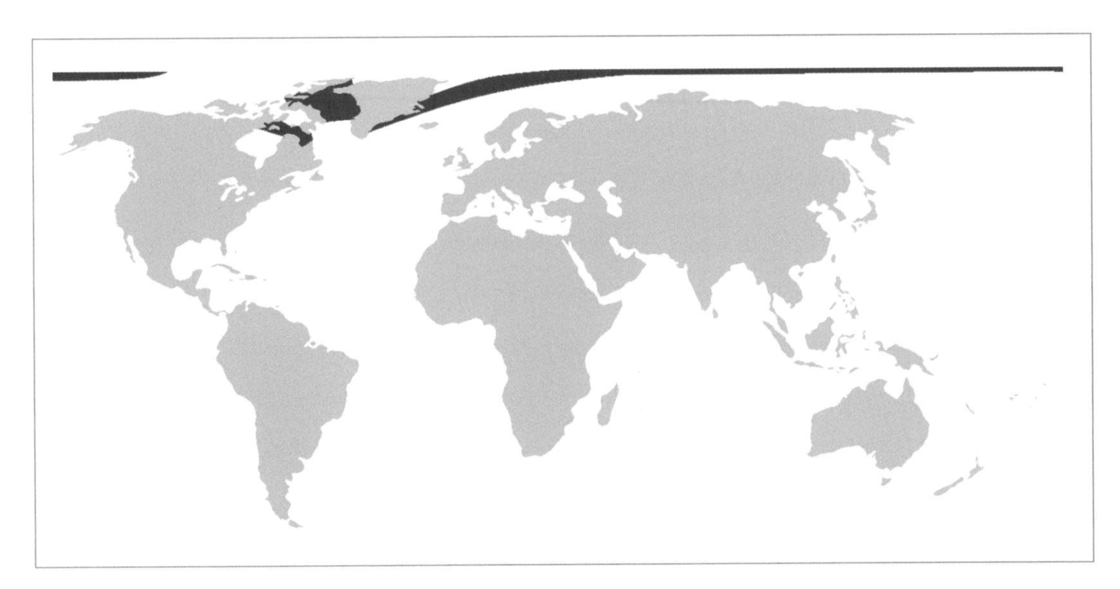

The present range of the narwhal (Illustration: Wikipedia Commons)

driven many species to extinction, yet many feel longing when discussing the dodo, the Tasmanian wolf or other extinct species equally exotic in distribution and specialization as the high arctic species. Some may even feel guilt on behalf of our predecessors. Extinction means irreversible change. If one has caused an irreversible change, one has something to feel culpable for.

The earth is complex and its processes are intertwined. Many of them remain uncertain and poorly understood, and thus establishing the origin of a particular change will for all practical purposes always involve elements of uncertainty. Should we take action when we are not absolutely sure that the action is the correct one for remedying something that has been identified as possibly negative? The precautionary principle states that if an action or policy is *suspected* of being potentially harmful to the public or to the environment, in the absence of scientific consensus that the action or policy *is* harmful, the burden of proof that it is *not* harmful falls on those taking the action. Shall we wait to act until we are *absolutely* certain there is a problem? Shall we prohibit all activities that have not been *proven* to be benevolent towards the environment? Both arguments, taken to the extreme, lead to inaction if elimination of *all* uncertainty is made a prerequisite for any decision. Inaction in a particular issue may be a desirable result in the eyes of some, but it is not a responsible way of tackling the dilemmas of humanity.

The basic needs, the ingredients for a dignified human life – food, health, and a safe home – require agriculture, medical industry, and housing. These in turn require space, chemicals, and materials. From this follows a need for transportation and energy. None of this can be achieved without altering natural ecosystems, creating synthetic chemicals and materials, mining, forestry, industry and power plants.

Do we have a responsibility to minimize how the actions taken to meet basic needs create negative effects for others? Should the fulfillment of basic needs be exempt from scrutiny of methods and consequences? How do we handle all the desires we humans tend to feel once our basic needs are fulfilled? Should such stimulus and enjoyment, which is a privilege restricted to the affluent in the world, be subject to higher scrutiny than the actions related to satisfying the basic needs?

Do we have a responsibility to leave some (or all) of the earth's systems untouched? Do we have a responsibility to preserve natural creation for future generations? The world and humans will continue to exist without the polar bear or the narwhal, but if we can avoid eliminating a species, do we not have a responsibility to do so? Should we feel responsibility for the wolf, the narwhal, the natural ecosystems, and the future of the Arctic?

Dilemmas arise when one person's actions impose an undesired change in someone else's ability to pursue their happiness, or when the actions have consequences for the environment, especially if these consequences are irreversible. Do the positive effects outweigh the negative? How shall we handle the fact that the positive and negative effects can be separated in time and space? How shall we handle the fact that the evaluation of both positive and negative contains uncertainty? Are the positive and the negative distributed fairly?

Who is to evaluate what is positive or negative? Who is to judge which effects outweigh each other?

Uncertainty about what risks we are taking does not liberate us from responsibility for the future. Ignoring the future can of course be part of an egoistic lifestyle in the present. But most of us find meaning in life through our children and friends and wish to bestow on them opportunities to fulfill themselves that are at least as good as the opportunities that were given to us. From such normative goals follow ethical consequences. There is an obligation to question an action when that action *to the best of our knowledge* jeopardizes environment and biodiversity, thus depriving the future of opportunities. People handle uncertainty every day without being paralyzed by inaction; we must be honest about uncertainties, and then build an understanding of what the best judgment is despite the uncertainties.

The disconnection between actions and their negative consequences in time and space makes us unable to foresee or grasp the consequences of our individual actions. This is also problematic in the reverse. The benefits of actions taken today (to counteract climate change) will probably not be experienced within the lifetimes of many who are asked to take the actions. The actions may not even be proven beneficial on that time scale. It can then be tempting to choose inaction based on economic arguments or convenience. What difference does the individual's decisions make – or even the choices of a region or a single country? Why should we take action if we cannot reap the benefits ourselves or even see benefits for others in our lifetimes? Decreasing our impact on earth's radiative balance can appear a daunting and demoralizing task. We must seek strength within ourselves and consider what the alternative is; do we wish to be remembered as those who did not even try?

Rewards, responsibility, hardship, and guilt cannot be fairly distributed by any metric since we in the foreseeable future will still struggle with identifying many of the connections. Endless quarrels about how to achieve fair distribution will only lead to exhaustion and, at best, inaction, and at worst, hostility and conflict. Fairness is a virtue; but it is necessary to take responsibility for the future now, and neither inaction nor conflicts are endowments to be handed down to coming generations with any pride.

The future will contain change: natural change and human-induced change. Even if we stop all emissions this minute, we have already perturbed the system such that there will be decades of continued climate change. Adaptation to human-induced climate change must thus be part of our future, whether we like it or not. Technological solutions to decrease emissions must be part of the future and we need innovation and willingness to try new solutions. But we must also look into our own conscience; innovations and higher efficiency do not mean we can relax. There has been extraordinary technical development, innovation, and increased energy efficiency during the past century – but our emissions have continued to increase. Emissions have increased due to population growth, but also because we all have increased our use of energy. It is tempting to utilize innovations not to alleviate our footprint on the planet, but to increase our comfort and convenience. There is a need to discuss how we utilize innovations such that we indeed give the Arctic and the world the future opportunities they deserve.

Innovations and new technology will always entail risks and uncertainty until they have been tested, proven and established. There are profound dilemmas related to testing new methods where the risks of testing in populated areas versus testing in pristine areas are pitched against each other. Trust – and distrust – in technology is widespread in many societies. Both the trust and the distrust sometimes have irrational dimensions, but ultimately all technology is designed and maintained by human beings. We must be humble: human error will always exist. The Arctic is fragile and unique. The Arctic should be exposed to a minimum of unproven technology and risks that arise from human weaknesses.

The changing Arctic is eagerly looked upon as giving new opportunities with easier access to oil and gas resources, mining, shipping, fisheries, and tourism. Must the affluent world utilize every opportunity, even when it increases pressure on an already beleaguered

Arctic? Maybe prudence is wise when technologies and consequences are highly uncertain in the Arctic. Could the mineral and fossil fuel resources in the Arctic be saved for more needy times, to relieve the area of further strain in the present troubled times?

Which human-induced changes are most important? The question is in many ways destructive. All changes are important to evaluate and weigh. Human-induced climate change is clearly one of the most serious perturbations in the Arctic, but it must not make us forget all the other drastic changes that humans are imposing on each other and the environment. Conversely the other changes must not be used as an alibi to downplay climate change. If we spend much effort on identifying the champion amongst changes, less effort is available to look for solutions, remedies and ways forward into a safer and sounder future.

What *should* we do, what *must* we do, and what *can* we do?

Decisions with far-reaching and profound influences on many people quickly become unpleasant if the people are not convinced by the motivations for the action. This makes inaction tempting as a refuge for many decision-makers. Knowledge and spreading of knowledge are the keys to making people willing to pursue difficult tasks.

Knowledge is the key to the future. Our knowledge is ever increasing but will in all likelihood remain incomplete. We will always seek more knowledge, but weaknesses in our present knowledge of the causes of climate change must not become an excuse for inaction or procrastination. The risks are so large that we must take action based on the best judgment we can muster with the best knowledge available today. As new knowledge emerges we must reevaluate earlier decisions. Without a doubt some decisions and old facts will be proved wrong as new knowledge emerges; there will be failures with new technologies and innovations, and there will be new negative consequences that appear as we make decisions and try new pathways. But then we must revisit the issues assiduously, without tiring, to bring ourselves back on a constructive and positive course. If we fear failure, we will never succeed. If we do not try because we think it is too difficult, we will never succeed.

Knowledge can be used for good and bad. Bertrand Russell struggled with finding a stringent and logical definition of what is right or wrong. A famous consideration of his is: "I cannot see how to refute the arguments for the subjectivity of ethical values, but I find myself incapable of believing that all that is wrong with wanton cruelty is that I don't like it". Establishing what is good or bad cannot be done inside a single brain through the principles of modern scientific reasoning or mere induction. It will require many brains in ethical dialogue and profound thinking about subjective matters. Democratic processes, remembering historic examples that must not be repeated, in an open and tolerant dialogue where people are invited to participate: this is a powerful way of identifying the core ethical values we wish to fulfill. Holding such ethical values high is the way forward and helps ensure that knowledge will be used for the good. It is a responsibility to capture the dialogue and guide it toward core values and not allow it to decompose into trivialities of rudimentary definitions of freedom. Dignified lives and basic human rights for all are more important than cheap energy, the latest fashions or gadgets for personal entertainment.

We need dialogue, debate, and democratic institutions so that the blessings of knowledge and spreading of knowledge are distributed fairly. Education is a birthright that we must fulfill for all young people. But in doing so, we must acquaint them with ethical dialogue and debate early in their education to counteract trends of complacency. Education, critical thinking, respect for different views, and a constructive use of knowledge to actively find the best solutions, are the keys to sustainable development – and of paramount importance for preserving democratic institutions. Education empowers people to go beyond their basic needs and enrich themselves and the cultures they live in through the abilities and opportunities that education gives them to influence their own futures.

We need dialogue where industry, governments, and individuals look for common goals. For an individual, the opportunities to influence and the ability to make choices are restricted by what industry is producing, what governments are promoting, and historic circumstances. The individual can only go so far. We must not impose guilt on individuals by demanding they make choices that are impossible within the society they are faithful contributors to. Taking responsibility for the future is not done by pointing fingers at individuals, sectors, or politicians. Climate change is a problem that requires the full commitment of educated and positive civilizations where all sectors, private and public, are devoted towards creating new alternatives and solutions.

The narwhal, the polar bear, the ivory gull, and all other arctic species are important. What is happening in the Arctic is important to make known – not to paint a picture of hopelessness and catastrophe, but a picture of beauty and hope. With education, knowledge, and dialogue, we can find inspiration for innovative ways of offering dignified lives to all human beings, whilst preserving the unique environments, and importantly, through use of the emerging and new technologies that we thus have inspired, also continue building human prosperity. The Arctic is a unique gem that belongs to everyone, and it is everyone's responsibility to pass it on to coming generations, as a gem.

Narwhals at the ice edge (Photo: Wikipedia Commons)

Notes

1 http://www.state.gov/secretary/remarks/2014/02/221704.htm
2 ACIA report *Impacts of a warming Arctic*, Cambridge University Press, 2004, p. 8
3 Hønneland, Geir: *Arctic challenges.* Kristiansand: Høyskoleforlaget, 2012, p. 18
4 Lecture at the conference "Global warming as an ethical dilemma", Longyearbyen, September 5, 2008.
5 Officially named the law on protection against tobacco-induced harm
6 As of 2004, restaurants, workplaces, institutions and meeting rooms must be completely free of tobacco smoking.
7 Norwegian radio, November 13, 2013
8 Why the Philippines wasn't ready for typhoon Haiyan, *Washington Post,* November 11, 2013
9 *Dagbladet*, October 28, 2012
10 *Dagbladet*, January 3, 2014
11 From the TV program *Watchers in the Arctic (Voktere i Arktis)*, NRK, 2002
12 *Six Degrees*, Mark Lynas, 2007
13 United Nations Environment Programme, www.unep.org
14 World Environment Day, 2007, introductory comments
15 Blog from Fram Strait, NRK.no http://www.nrk.no/nordnytt/bli-med-nordover_-1.5645790
16 Blog from Fram Strait, NRK.no http://www.nrk.no/nordnytt/ved-veis-ende-1.5839557
17 http://www.nrk.no/nordnytt/--ekstremt-lite-kunnskap-1.6667239
18 http://www.youtube.com/watch?v=ztz3ZdPbdKo
19 www.nrk.no/nordnytt/arktis-pa-vippepunkt--eller-over_-1.6682606
20 http://news.nationalgeographic.com/news/2007/12/071212-AP-arctic-melt.html
21 IPCC Report, 2014, www.ipcc.ch
22 http://www.climatechange2013.org/images/uploads/WG1AR5_Headlines.pdf
23 NRK, www.nrk.no/nordnytt/mindre-is-bekymrer-forskere-1.7288001
24 *Six Degrees*, Mark Lynas, 2007

25 Wikipedia entry on Garrett James Hardin
26 From Norsk Natur 1/74
27 Researchers have found the reason for the sudden cold shock in Europe (in Norwegian), *Aftenposten*, November 29, 2012
28 http://natgeotv.com/no/six_degrees/about
29 NSIDC, National Snow and Ice Data Center, August 27, 2012
30 The Climate Reality Project
31 Hotel Tropical. Homeland of nights, Oslo 2003
32 From Torgrim Titlestad: Peder Furubotn 1890–1938. Oslo 1975, p. 19
33 The Road to Wigan Pier, London 1937
34 Solheim, from Torgrim Titlestad, ibid
35 Pyramiden. Portrait of an abandoned utopia. With photographs by Siri Hermansen, Oslo 2007
36 "Das Licht der Arbeit ist ein schönes Licht, das aber nur dann wirklich schön leuchtet, wenn es von noch einem anderen Licht erleuchtet wird." Vermischte Bemerkungen, (1937), Frankfurt a. M. 1977 p. 56
37 http://www.imr.no/temasider/fisk/torsk/polartorsk/status_rad_og_fiskeri/status_rad_og_fiskeri/nb-no
38 Climate Change 2013: The Physical Science Basis, www.ipcc.ch/report/ar5/wg1/
39 Climate Change in Norway. Alfsen KH, Hessen DO, Jansen E. Universitetsforlaget 2013
40 Norway's "moon landing" concept alludes to the ambitious US Space Program launched by John F. Kennedy in 1961 with the words "this nation should commit itself to achieving the goal, before the decade is out, of landing a man on the moon and returning him safely to the earth". The US Space Program also had significant technological spin-off effects.
41 IPCC AR5, Report from WGII: Impacts, Vulnerability and Adaptation
42 IPCC AR5, Report from WGII: Impacts, Vulnerability and Adaptation, Chapter 7. Food Security and Food Production Systems
43 United Nations general assembly resolution a/RES/64/197
44 Westengen, Jeppson & Guarino (2013) Plos One, e64146
45 Svalbard receives 1000 more hours of light per year than Mecca [in Norwegian], Jon Børre Ørbæk, Ottar No. 3, 2005
46 http://www.riksantikvaren.no/filestore/SusanBarr.pdf
47 https://www.etikkom.no/en/vart-arbeid/hvem-er-vi/komite-for-samfunnsvitenskap-og-humaniora/
48 https://www.etikkom.no/forskningsetiske-retningslinjer/Etiske-retningslinjer-for-forskning-pa-menneskelige-levninger/
49 Sætersdal, G. and Loeng, H. 1987. Ecological adaption of reproduction in Northeast Arctic Cod. Fisheries Research 5: 253–270
50 International Geosphere Biosphere Program
51 National Oceanic and Atmospheric Administration, USA

52 Parts per million
53 https://unfccc.int/bodies/body/6383.php
54 https://unfccc.int/kyoto_protocol/items/2830.php
55 The United Nations program on reducing emissions from deforestation and forest degradation in developing countries, see www.un-redd.org
56 http://forskning.no/meninger/kronikk/2013/12/klimalosningene-kan-vaere-en-del-av-problemet
57 As presented in the daily newspaper Bergens Tidende, March 14, 2014
58 Read more at www.ny-aalesundsymposium.no
59 http://www.ices.dk/news-and-events/news-archive/press-releases/Pages/Press-release---Exploitation-of-fish-stocks-has-declined-significantly-during-the-last-decade.aspx
60 Jakobsen, T. and Ozhign, V. 2011. The Barents Sea Ecosystem. Tapir Academic Press, Trondheim, 825 pp
61 Gullestad, P., Aglen, A., Bjordal, Å., Blom, G., Johansen, S., Krog, J., Misund, O.A. and Røttingen, I. 2014 Changing attitudes 1970–2012: evolution of the Norwegian management framework to prevent overfishing and to secure long-term sustainability. ICES Journal of Marine Science (2014), 71 (2), 173–182. doi: 10.1093/icesjms/fst094
62 Alun Anderson 2009. After the Ice. Virgin Books, London, 298 pp
63 http://www.sysselmannen.no/Nyhetsarkiv/Forslag-til-forvaltningsplan-for-Ost-Svalbard/
64 http://www.sysselmannen.no/Toppmeny/Om-Sysselmannen/Sysselmannens-oppgaver/Miljovern/Forvaltning-av-verneomrader/vest-spitsbergen/